Palgrave Practical Guides in Communication

Palgrave Practical Guides in Communication provides accessible introductions to all key topics in the field, with an emphasis on interpersonal, intercultural, and organizational communication, as well as public speaking and rhetoric. The series includes both textbooks geared towards specific courses taught in communication studies programs across the globe, and practical guides geared toward both advanced students and practitioners. Volumes in the series emphasize not just the what, but the how: they help readers improve specific communication skills and practices. In order to do that, volumes in the series include a variety of exercises, activities, and other pedagogical features, some of which can be multimodal (e.g., video and/or audio). Informed by cutting-edge research by leading scholars, volumes in the series help readers become more effective communicators in today's rapidly changing world.

David Dannenfelser

The Art of Effective Science Communication

A Performer's Guide to Public Speaking

David Dannenfelser
Theater
Rutgers University
New Brunswick, NJ, USA

ISSN 3004-9032   ISSN 3004-9040   (electronic)
Palgrave Practical Guides in Communication
ISBN 978-3-031-57029-2   ISBN 978-3-031-57030-8   (eBook)
https://doi.org/10.1007/978-3-031-57030-8

This Palgrave Macmillan imprint is published by the registered company Springer Nature Switzerland AG
The registered company address is: Gewerbestrasse 11, 6330 Cham, Switzerland

Paper in this product is recyclable.

For my father, Lawrence Dannenfelser, a man who sought to connect to everyone through humor and kindness. I owe him for my point of view of the world, imperfect as it is and as he was.
For my family, my support system through it all.
And for my friend and colleague, Nick Ponzio, a master communicator and mentor to me and countless others. You are missed.

PREFACE

This is not a textbook. I just want to be clear about that. And because this is not a textbook, the charts, graphs, tables, figures, and even citations ordinarily associated with the textbook format will be kept to a minimum. Credit is given where credit is due, but most of what you will read in these pages is there because of my direct experience. And so rather than following an academic approach, I have elected to create a more conversational tone for this book. I will be speaking in the first person, relaying stories and accounts of events that have influenced me personally. This is by design, of course, because this is a book about communication and not merely an exercise in scholarship. What is traditionally accepted as proper procedures in academia will be challenged and at times characterized as possible roadblocks to connections between human beings attempting to talk about human things. In other words, not so many charts and tables.

Now I know this will be frustrating for some of the more scientifically oriented readers. Please understand that I respect the work of scientists and I acknowledge the importance of data, but this book is not about that. I also understand that there are many kinds of audiences who have more or less of a connection to the work of scientists. I will talk about that in later chapters, but it is worth mentioning here that the approach of this book is concerned with the communicator's efforts to reach audiences of any kind. Knowing your audience is important, but understanding what is common to all of us and how we share the same curiosities and thirst for information as a part of our nature as human beings is more to the point of this book. Data and evidence are essential, but not the focus here. These are a part of your preparation for your work as a scientist, but in my view, these can sometimes be a barrier to your efforts to communicate that work. The data should be available for any audience member to inspect, but later after the interest in your work has been sparked by the way you talk about that work with clarity and passion.

To that end, I will use empirical examples from two distinct groups when describing certain principles and sharing certain experiences. Theater artists and scientists will be mentioned often and the shared goals that support the relationship that exists between these will be pointed out frequently. I will also mention the medical community at times just to mix it up a bit. If you don't identify with any of these groups, don't worry, there is plenty here for you too. Because, although I have focused on these groups, the ideas and principles expressed in this book can be accessed and made use of by anyone who wishes to improve upon their communication skills. The information found in these pages can be significant for anyone with a desire to connect and be heard.

Why then do I spend so much time with theater artists and scientists? There is a reason for this. The very idea for this book grew out of my work as a part of a team of educators charged with the task of teaching scientists how to talk about their work in a conversational and accessible way. The Communicating Science Course I teach along with several other professionals from the science disciplines has been drawing students at Rutgers University to this new approach to public speaking for some time now. They, like many of you I suspect, have been searching for a way to make their work known to more than just their colleagues.

So, who then is the audience for this book? You and anyone like you with a desire to connect.

New Brunswick, NJ, USA   David Dannenfelser

ACKNOWLEDGMENTS

I would like to thank the following people for their invaluable input to this book and for their continuing dedication to the art of effective science communication.

Janet Alder, PhD
Associate Professor Neuroscience & Cell Biology
Assistant Dean School of Graduate Studies
Rutgers University, The State University of New Jersey

Helen Farmer, MFA
Assistant Professor of Theater
Molly/Cap 21 College

Jamie Grady, MS
Arts Administrator
Executive Director for 2/2 Arts Management
Co-owner and Project Manager
J. Hendricks Homes, LLC

Joe Giardina, Program Director
Director Of The School For Performing Arts Bronx House

Kevin Kittle, Professor Emeritus
Mason Gross School of the Arts
Rutgers University

Robert C. Like, MD, MS
Emeritus Professor of Family Medicine and Community Health
Rutgers Robert Wood Johnson Medical School

Dennis F. Mangan, PhD
Director, Chalk Talk Science Project
Former NIH Senior Advisor
Former Associate Dean for Research, University of Southern California

Mary L. Nucci, MS, PhD
Assistant Dean, Campus Engagement
Assistant Professor, Department of Human Ecology Rutgers, The State
University of New Jersey

Carol A. Terregino, MD
Senior Associate Dean for Education and Academic Affairs
Rutgers Robert Wood Johnson Medical School

CONTENTS

CHAPTER 1

Introduction

Abstract Theater is about life and the quest to understand it. As artists, we seek connection to our fellow human beings through our expressions in the form. As it turns out scientists and just about everyone else are interested in the same thing. Some of us are more introverted than others and maybe a bit awkward in our attempts to connect, but the desire is still there. That was the motivation behind this book, to suggest some techniques by which anyone interested in expressing themselves and being understood by others could find a way to simply connect.

Keywords Role taking • Drama Therapy • Working in-role • Theater games • Effective communication

I am an introvert and I work in theater. But before becoming a theater professional and educator, I was a special education teacher. I know, not the usual path to the art form, but what has always been true of my work has been that teaching and education have been at the center of it. My early experiences in theater were always influenced by my background in education. Though I had been fascinated by the theater arts early on, I couldn't commit to a life as an actor. I pursued a teaching career in special education not sure of who I was but knowing that teaching would somehow figure into my future. Later when fate or whatever had brought me together with an old friend who worked in theater, I began the process of merging my interests in the art form with my work as an educator.

These twists in life can be organic as they were in my case. I was astounded to discover that I could pursue both of my passions through a program at NYU. It was there that I found I could major in Educational Theater. There it was, the encapsulation of my interests and possibly even a new career path. Truthfully, I never felt my work as a special education teacher was a good fit for me. I don't regret a minute of it though because it led me to the work I do now albeit in a roundabout way. It seems that was to be the way I would come upon every new discovery about myself and my work. I did things the hard way it's true, but I also found that although I followed a path that was undetermined at times, it often led me to somewhere significant and ultimately to the place I needed to be.

I discovered long ago that acting was not going to be my primary focus as a theater artist. I did grace the stage early on and even then, was aware of my preference for the creative activities of writing and directing. When I did appear as an actor, it was mostly in an educational setting using my performance to teach or explore a social issue as a member of an educational improvisation company ironically named, *Good Clean Fun*. My introduction to that company started when they were still working in the bars and comedy clubs of New York and New Jersey doing audience suggested, production improv. I had a turn at that, but never really took to it the way my fellow improvisors did. I preferred the longer form structure of scene work. It was when I became the education director for the company, and we moved out of the bars and into the classroom that my work really began.

As I've said, I'm an introvert and so my comfort level with performance was always a challenge for me. I was much more content to sit in a room alone and write plays although I didn't know that about myself until later when I did exactly that. So, at the beginning, performance was my only means of expression in the arts. Though it was difficult, I did discover something about acting that allowed me to be on the stage and to enjoy myself. I learned about character and the pursuit of a clear and specific objective and of how these things created a structure for my performances and even gave me permission to be on the stage, something I didn't believe I had at the time. The use of role taking in my work was a significant tool that allowed me to do things I didn't feel I could do or had a right to do. It wasn't until very much later that I came to study this concept as a graduate student at NYU. There I was exposed to the concepts of Drama Therapy through a course on the same taught by Robert Landy, an expert in the field.

Working in-role was just one of the techniques I would learn about and later apply to my own career as an educator. As a matter of fact, it was often the case that I would discover what a technique was called after having practiced it for years in the improv company. Giving official names to things I had been doing was my backwards process for growing in my understanding of the power of the art forms of theater. It wasn't until I met my friends in *Good Clean Fun* and then went to NYU that I really started to see how theater was more than just about putting on shows for entertainment. This was an ancient and effective means to communicate with others.

The power of connecting to other human beings using drama was a concept that I had to grow into through experience. Though it was a lot of fun to get together with fellow company members and play theater games, it wasn't until we started to use the structure of those games to develop an effective means of communicating with others that the awesome nature of the art form became apparent to me.

And so, here I am today using my experience and the skills I've developed as a theater professional to teach scientists and others about effective communication. I still write and direct plays and I still teach acting students and others about theater, but the boundaries that used to define those roles are a little more flexible for me now. The journey to understanding the connections between the disciplines of art and science has been a long one and very rewarding. I am hoping to continue to find common ground between these different but related fields of study.

Theater is about life and the quest to understand it. As artists we seek connection to our fellow human beings through our expressions in the form. As it turns out scientists and just about everyone else are interested in the same thing. Some of us are more introverted than others and maybe a bit awkward in our attempts to connect, but the desire is still there. That was the motivation behind this book, to suggest some techniques by which anyone interested in expressing themselves and being understood by others could find a way to simply connect.

PART I

Theory

CHAPTER 2

Process Versus Result Orientation

Abstract This first chapter lays the groundwork for the approach to public speaking that will be discussed in the book. There are some similarities and many differences between the traditional approaches to public speaking and my view of the process. Comparisons to what has been acceptable practice in the past to new ways of looking at these things are outlined and explored in this chapter.

Keywords Process • Result orientation

A Word of Warning

To all you scientists and non-actors out there, I'm going to be talking about acting for a while. I'm still talking to all of you, but I'll be using mostly actors and other artist types as my examples for the points I want to make. Please understand that I'm still thinking of you when I do this and I will also include you along the way, but this book is about how the actor's process can be understood and accessed for other purposes and by other people. That means you too, I hope. Please bear with me. I promise I'll bring it back around to all concerned.

So What Is Effective Communication?

There are many answers to this question depending upon who you are talking to and in what context. I hope to make a distinction between the more traditional approaches to effective communication and my own approach that is influenced by the work of theater artists. The use of improvisation and acting techniques to connect to an audience is time tested. The theater arts have

D. Dannenfelser, *The Art of Effective Science Communication*, Palgrave Practical Guides in Communication, https://doi.org/10.1007/978-3-031-57030-8_2

addressed the desire to communicate and share in a sense of community with fellow human beings for a very long time. To be an effective communicator is to be heard and understood by others. There are traditional approaches to this goal and there are others like those expressed in this book. In the end you will have to choose what works for you. Let's begin by looking at some of the more traditional approaches and then consider some others.

What Has Been and What Could Be

I recently came across a description for a Public Speaking course in which I found a statement about the many important things a student should learn to become a better public speaker. Some of these included presenting themselves and their material clearly, confidently, and persuasively. The students could also expect to learn about techniques for overcoming stage fright and for developing clear enunciation by finding their natural unaffected vocal register and varying tone and intonation to hold audience interest. Pacing, moving with assurance and purpose, and using appropriate gestures along with eye contact are also described as techniques for better public speaking. All of this will apparently help a student to change behaviors that keep them from being better communicators and instead help them to maximize their persuasiveness.

There is much to be considered in a course like this although the point of view of the approach as described may be somewhat misguided. Not surprising as these tenets have been celebrated and endorsed as good communication for a very long time. Many of these ideas have helped people to become better public speakers and so should not be discounted out of hand, but there may be another way of supporting those who wish to become the best communicators they can be within their chosen field. It is in that spirit that we look at these same ideas but take a decidedly different approach to achieving some of the same goals described above. In my view, good communication is guided by a study of the principles found in the training for acting and improvisation, not Public Speaking.

So how then does one present themselves more confidently or persuasively? Likewise, how does the student of public speaking overcome stage fright or develop clear enunciation? Is it by finding their natural unaffected vocal register and varying tone and intonation to hold audience interest? How do you do that? How do you work on your pacing and moving with assurance and purpose and using appropriate gestures along with eye contact? Okay, eye contact is easily made, but will just making eye contact with your audience make you move with more assurance and purpose? And what is an appropriate gesture? How do you know if your gestures are appropriate or not? You could assume that these questions will be answered when you take the course, but will they, I mean will they really?

I ask that because the original premise for a course such as this is based on a belief that the result is the most important thing for the student to focus upon when speaking to an audience. How you sound, how you move or even what you look like are very important considerations according to the instructors of traditional public speaking courses. This result orientation is at the core of many such courses and is also the tragic flaw contained in this approach as I see it. A student who focuses on these things is already taking on a self-conscious point of view before they've even set foot onto a stage.

Imagine if you will that you are watching a sporting event on television, let's say basketball. As you follow the action, you notice something strange happening with one of the players. He is dribbling down the court using all his skills to play the game, but something else is happening as well. As he dribbles the basketball with one hand, he holds a mirror in the other and proceeds down court all the time watching himself in the mirror to see how he is doing. Ridiculous you may say, and you would be right. Besides creating a distraction for himself that will almost certainly cause him to lose possession of the ball, his focus is not on the task at hand, namely weaving his way through a sea of his opponents and possibly scoring a basket.

This is an absurd example of course, but how much more absurd is it for you to watch and judge your own performance as you speak to an audience about your work? In this analogy, you are the one who loses focus and takes your eye off the ball to self-consciously watch yourself and of course find yourself to be lacking in some way. Your focus is on the result of what you are doing and not on the doing itself. This is a recipe for failure, but also an essential part of the experience for far too many people who have decided to take that traditional public speaking course. These people have a sincere wish to get better at what many believe is a daunting task, talking to a group of people about the thing you love, your science research. What they find is that they are being encouraged to hold up a mirror to their efforts rather than getting busy doing the work.

Actors sometimes must share a space with other artists. Dance programs and acting programs have something in common besides being parts of the performing arts. Both are too often underfunded because they are a part of what our society considers to be fringy or soft skilled pursuits. Those of you in the sciences may experience some of these challenges, but I must tell you, you don't know the half of it. The arts are in a constant struggle to survive, but that is the reality of the circumstance and so an actor will sometimes have to rehearse in a dance studio and vice versa.

For the dancer, form is an essential part of training. The aesthetic considerations of how you move are key to your mastery of the form. The fundamentals are concerned with learning certain positions and in executing those positions perfectly. You must therefore watch yourself in a mirror as you learn to Assemblée, Plié, or Pirouette. This is a part of the dancer's training. When that same dancer is in performance the mirror is nowhere to be found on the stage. It stays where it belongs in the rehearsal room. The focus on stage is simply to dance.

Actors too pay some attention to form and movement as a part of their training, but the actor has action as the first concern. Doing is first. Watching yourself do is to be avoided as much as possible. There may be some consideration for movement, but very little consideration for the mirror for the actor. And so, if an actor needs to share a space for rehearsal with dancers, the first thing they do is pull the curtain over that mirror. If the actor isn't in a class on movement, they understand that to watch how you do what you do is to focus on the result. Like the dancer, when the actor is in performance, they leave the mirror behind. In either case there is an understanding that you don't bring your homework onto the stage. An actor understands that when they are in the moment of the staged reality, they are only interested in knowing who I am, where I am and most importantly what do I want. If I trust my training and leave my homework in the rehearsal room, my form will take care of itself. The dancer too must at some point trust in their training and forget about all the forms to focus on the dancing. There is a moment whereby everyone must leap into doing and trust that the training is there, but now transformed into doing.

To repeat, the focus for actors is on doing. Actors make choices based on action rather than on state of being. To refer to that traditional public speaking course, just wanting to be more persuasive is not enough for the actor. To simply *be* anything is inactive from the actor's point of view. The focus must be on the action that may or may not lead a person to persuasiveness. Though both the traditional and the actor's approaches are concerned with goals, the actor's training shifts to the process by which you achieve your goals and not on the goals themselves. The actor believes in doing. The end goal or result will happen only if the actor is busy pursuing an objective that has that goal in mind but doesn't force the artist into the indication of that goal.

In short, the actor understands that goals are the end-product of an active pursuit of something specific that is wanted from a partner in the moment. This in turn creates behavior in the actor which is connected to that specific want, desire, or objective. In the end the actor understands that they have no control over the outcome, i.e., being persuasive or not being persuasive, but rather that the outcome depends upon a specific devotion to the action or verb that may get you there. More specifics on this are covered in a later chapter, but for now it is worth noting the major differences that exist between the traditional and the actor's approach to public speaking. Traditional public speaking training is concerned with results. The actor's approach is more concerned with *how* you get those results.

There are a few more things to consider as we explore this approach to better communication. In the chapters that follow we will look at some foundational beliefs that I hope will lead a student to have a positive experience of public speaking. We will look at the nature of spontaneity and how it is possible for it to be a part of the speech you share with an audience by letting go of the expected and being present in the moment. We will also discuss the concept of goodwill and how it is important to believe in the positive nature of the circumstance to see yourself as accepted and even encouraged by an audience that wants to hear what you have to say. And we will speak about forgiving yourself when things don't go as planned. Learning to let go of the inner critic and moving on are ideas that are easy to talk about, but difficult to live through when you take a risk and put yourself out there in a very public way.

Which leads us to probably the most important aspect of the actor's approach to public speaking, the concept of trust, trust in the process, trust in the reality of the circumstance, and trust in yourself. The student of public speaking must be brave and trust in the notion that they have a right to say their piece no matter what anyone else thinks about it. These are the things I want to talk about in the following pages. If I have sufficiently piqued your interest and you want to learn more about the *how* of these things, please read on.

SPONTANEITY: LETTING GO OF WHAT IS EXPECTED AND STAYING IN THE MOMENT

As mentioned in the introduction, years ago I served as the Education Director for an educational improvisation company, *Good Clean Fun*. Our work involved developing scenes for discussion with school groups and community organizations. We followed the teachings of Viola Spolin as outlined in her book, *Improvisation for the Theater* to create our own original short scenes on social issues. Our rehearsals were as much about playing as they were about developing material for the scenes. For us these two things were intricately linked. Playing was essential to our process and lead us to eventually write many practical pieces of theatre for our performances and workshops.

There was a joke well known to the members of our group that referred to a fictional posting for the meeting of an improv group. The posting read "Improv auditions tonight, prepare something spontaneous." The absurdity of this juxtaposing of ideas said a lot about the process we all had committed to as members of this company. The nature of improvisation would suggest that it is impossible to "prepare" for spontaneity, but there was also some truth in that phrase. Scenes were never written down. Beats were recorded. (More on Beats later.) The rest was in the moment. While the spontaneous moment is just that, there is much to be done in preparing for that moment's arrival. You carefully create the circumstance and the particulars surrounding it such as the who, the what, and the where. You then remain open to the moment when it arrives.

You do your preparation and then leave yourself alone and let the magic happen. If you've ever seen a magician at work, you may know that what you see, the trick is a result of endless hours of rehearsal and the development of skills. But when the actual illusion is achieved the magician is in the moment and letting his training and skill do the work while allowing for the magic to be revealed. As audience members we are not aware of the mechanics of the trick. If we were, it would cease to be a trick therefore robbing us of the magical moment in our experience. We can spend a lot of time questioning the how afterwards, but when witnessing the trick our experience is of wonder at the magic.

APPROVAL AND DISAPPROVAL: FORGIVING YOURSELF AND MOVING ON

Perhaps one of the most difficult things for anyone to do is to let go of judgment. I'm talking about the judgment we receive from others when we take a risk and attempt to make a public statement about our work, beliefs, or values. Everyone knows how hard it can be to take that risk, to put yourself out there and to encounter possible challenges or even ridicule for what you have to say. A scientist presenting findings or advancing theories is expected to have prepared evidence for any claims they may make. That is as it should be, but it is not unusual according to some scientists I have worked with to be shouted down at a conference because someone in the audience disputes your findings. It seems to be a bit of a tradition to do so. The rudeness of such an occurrence aside how is one to deal with this possibility?

The first order of business is to make sure you have followed the rules of best practices and to be sure that your work is grounded solidly in scientific methods before you venture out to speak about it publicly. This is a given of course but it should be stated clearly anyway that preparation is a huge part of any work you do as a scientist or as an artist for that matter. An actor prepares and so should the scientist before they speak publicly about their work. If you are prepared, then you will have an answer to a nay-sayer right there in the moment and answer you must. No challenge to a well-prepared sharing of information should go unanswered. As a scientist you must refer to the facts. It is therefore imperative that you have them before you share your voice. Being prepared is the best way to ward off any apprehension you may have about speaking up and it goes a long way towards creating the confidence in yourself that is necessary to do so.

But there is also something else. Beyond your readiness to speak about your work publicly, you must also believe in your *right* to say your piece. Some have spoken about imposter syndrome, that feeling you may have that you are not really the one to be telling others about your work. In the theater, an actor grapples with this same feeling at times. Am I really a good enough actor to have been chosen for this role? Do I have a right to be up on this stage in the place of another? Well, there is no other way to say this, yes you do. If you have put in the time and done the work to prepare for this moment, then the moment is yours. Take it and run with it.

I know this is easy to say, but it is also the truth. Our fears about approval and disapproval are baked into our psyches early on. A combination of an introverted nature and a disapproving atmosphere in which we may have been raised can make it very difficult to give ourselves permission to say what we mean and mean what we say. There is no formula for how you get over this. You simply must practice a healthy disregard for what others think of you or as Ray Charles is attributed as saying about his work as a musician, "I don't give a damn who like it or who don't." But simple ain't easy. When you consider that for many of us the harshest critic is ourselves then your work to overcome your apprehensions is that much harder to do. When an actor makes a mistake in the rehearsal process, we have a simple phrase we say to them, forgive yourself and move on.

Believing in the Concept of Goodwill

In my work as a theatre artist the concept of goodwill comes up often. Because the theatre experience is largely made up of a series of beginnings and endings, the actor, director, designers, technicians, and the playwright going from project to project and essentially starting over each time, it becomes important for each of these professionals to have a work ethic that allows for the transient nature of the world of theatre.

Each time an established work of drama or new play is mounted as a production, the team assembled is expected to form an ensemble, a cohesive group with shared interests, goals, or values. But people are different. This is particularly true of theatre artists. We pride ourselves on our individuality. We work hard to have our art stand out among the many offerings as unique. Even so, when the time comes to see that work realized on stage, commonality is the glue that holds the whole thing together. An understanding of certain truths held dear by the artist is essential for all to engage in a harmonious and functional working relationship with other professionals in the field. But how is this possible when you are surrounded by some very strong-willed and *interesting* people who call themselves artists? It comes down to the concept of goodwill.

Consciously or unconsciously, when a group of theatre artists come together to begin work on a project, they enter into an agreement with each other that includes in it the assumption of goodwill. When I say I want to create a thing of artistic relevance and beauty with my fellow artist, I am also saying that I will honor that unwritten code to accept and affirm everyone's right to be a part of that creative process. I assume goodwill on the part of everyone involved. I choose to believe that everyone is honestly working towards doing their best work to bring this thing into being, that which has never existed exactly in this form before. I am saying that I will see and hear you and though I may disagree, you will have your opportunity to make your case for your views.

This is a question of trust, but it is also a matter of practicality. When we make our assumption of goodwill, we are also saying that there is a means by which any argument or disagreement can be managed. Our default is always to the work, the reason after all for all to be here. We don't make our case for ourselves. We make it for the sake of the work. In that way, the work itself becomes the ultimate arbiter of any disagreement. We also take the temperature down a bit as we move away from the personal and closer to the professional.

That is not to say that the artistic workspace is wholly democratic in nature. The director is the director, and that person has the final say on any choices, but the choices must be heard. If in fact a person in the ensemble is simply voicing opposition to certain choices to call attention to themselves, that too is managed simply by questioning that person as to the relevance of their complaint to the work at hand. If there is found to be merit in the argument, then that is seriously considered. If the opposition can't be justified as to its efficacy to the work, the argument is settled there and then by pointing this out to the individual.

GOODWILL AND YOUR RELATIONSHIP TO THE AUDIENCE

How then does this concept relate to you and your efforts to speak publicly as a scientist? Very well I would say because if what you are doing is truly about the work and not about you, then your focus is on that work and not on yourself or any perceived slight to you personally. You have a job to do, a task to accomplish. Believe it or not, the audience is more often on your side than against you. If you are in fact successful in your sharing, then that audience gets what it came for that day. Who shows up to a forum to hear about a new scientific discovery to debunk that very same theory? Some people do make this a goal, but that is a sad occurrence for a minority of audience members.

When you're sharing some vital information about your work, the audience in attendance at your speech has entered into an agreement with you whether they are aware of this or not. The audience has decided consciously or unconsciously to give you the benefit of the doubt. They have agreed to enter the sublime and beautiful if unwritten practice of granting you the assumption of goodwill. They believe that your intentions are honorable, and they answer those intentions in the same vein. Why else would they bother to attend? For those few who are intent on sabotage, you must be settled with yourself that there is nothing you can do about it. Certain people can only feel validated in themselves by the destruction of others. This is not a healthy person and should not be considered to have any power over you and your goodwill attempt to share your work.

TRUST

And so, we come to trust. It is perhaps obvious to say that to achieve anything you must first believe and trust in your ability to do it. As we have been saying, it is also necessary for you to trust in the goodwill of others, to believe that people are on your side and want you to succeed. How do you do this, allow yourself to trust in yourself and in the goodwill of others? I have no idea. All I know is you must. Maybe it comes back to the idea of time and experience. If you put yourself out there enough times and find that ultimately you can survive anything that may happen, then maybe you will be able to put your trust in yourself, in others and most importantly in the process. That may be the answer. To put your personal fears and apprehensions aside and to look at the process as a means of building trust in yourself. That of course has not been the traditional approach in the past, but as I have said, we are considering new approaches that may challenge the traditional and hopefully improve upon it.

THE THRESHOLD OF TRUST

There is an exercise I do whenever I start to work with a new group. The exercise is simple. Everyone stands side by side at one end of the room. I stand at the other end of the room. By this time, I have learned each person's name through another exercise. One by one I call on each person to close their eyes and walk forward towards me. I also ask them to keep their eyes closed and to keep walking until I stop them by gently placing my hands on their shoulders. Simple? Yes, but simple ain't easy.

The fear that this exercise conjures up in some people is palpable. Everyone takes a turn, including me. Each time the group watches as one person takes those steps towards me, blind. There are many trust exercises, but this one is the simplest and I believe the most effective for a group who are new to this kind of process. To just walk forward, eyes closed and trust in someone to stop you may not sound like much of a challenge and to you who think it isn't I say, try it.

Why do I do this so early in our process together? Well, the answer is also simple. Trust as a concept can be very hard to fathom. A philosophical discussion about why trust is a necessary part of any creative endeavor may be of some use to people, but a discussion will only take you so far. Also, to talk about trust and intellectualize about it as a concept is an exercise in avoidance in my opinion. Our socially defended responses tell us that we should always be on guard. Self-preservation is a very real instinct, but it also calls upon us to be closed off from experience. To trust another out of hand is unnatural, but imperative for us to get beyond our defenses and out of our own heads. For us to risk anything, we must first have some degree of belief or trust in our ability to survive whatever challenge we may find before us. To just talk about this idea isn't enough, you must feel it.

And so, we engage in this physical experience whereby we are challenged to put our trust in another and in so doing learn to trust ourselves, our own instincts, and perceptions. Why do I call it a threshold then? If you ask someone to walk towards you with their eyes closed, you will see the moment exactly when their trust is being tested. You could if you wanted to draw a line on the floor where it happened. This is the threshold, the literal, physical distance I am willing to travel before my fear kicks in and tells me to preserve myself, protect myself from this stranger. This is my line; my threshold and I will not cross it without some hesitation.

I will often point this out to people in the moment. I will go to the very spot on the floor where you stopped trusting me and your defenses went up. My purpose here is to make that line of demarcation a point of departure. Here is as far as you are willing to go *now*. Let's see if that changes. I tell them I will revisit this exercise later in the process to see if people can take even just one more step beyond that line and take a risk to trust just a little bit more. Growth takes time. Building trust in others and in yourself can take a lifetime but work on this part of our nature to break through and increase our potential for creative experience is time well spent.

HUMAN NATURE AND SUCCESSFUL THINKING

The passage I shared with you at the start of this chapter is right in keeping with some very long-held beliefs about success in public speaking. Unfortunately, these emphasize character flaws as being the obstacles for many who wish to be more successful. There are some very famous books on the subject. In these the reader is offered instruction on how they should change their self-defeating nature.

Norman Vincent Peale's book, *The Power of Positive Thinking*, is a case in point. According to Peale, if you just think more positively you will be able to overcome your shortcomings. Your lack of confidence is a character flaw. But take heed, all you must do is stop acting in a certain way, be more positive and you can fix yourself and be more successful. Being a minister, Peale also puts a lot of emphasis on the spiritual and counsels the reader to turn to prayer for help with these negative tendencies. If you are religious, prayer may indeed be helpful to you, but not everyone is religious.

Peale also goes to great pains to equate spirituality with science saying, "Prayer power is a manifestation of energy. Just as there exist scientific techniques for the release of atomic energy, so are there scientific procedures for the release of spiritual energy through the mechanism of prayer." (Peale 1952) It is doubtful that the scientific community would not debate such a statement. Peale makes many such claims and often makes generalized references to famous scientists, psychologists, and health professionals, never citing anyone specifically. He casually refers to these ideas as "scientific spiritual practice" (Peale 1952).

His further assertions concerning prayer and its scientific uses come more to the actual point. In another generalized example, Peale states, "An illustration of a scientific use of prayer is the experience of two famous industrialists, whose names would be known to many readers were I permitted to mention them. Being schooled in scientific practice, they believe that in dealing with prayer as a phenomenon they should scrupulously follow the formulas outlined in the Bible, which they described as the textbook of spiritual science" (Peale 1952). Here again he equates the practice of spirituality with that of science, but now the intention is more clear. The bottom line here is the idea that spirituality as expressed through prayer is a kind of science that ultimately will lead one to success in business.

The capitalistic aspirations rooted in Peale's advice are numerous and share similar views with those expressed by Dale Carnegie, another example in the you are a failure because you believe you are a failure school of thought. Not a few aspiring titans of industry were given a copy of his book, *How to Win Friends and Influence People* back in the day. The premise here is again that you too can be successful if you follow a few simple principles and ignore your own essential nature.

Carnegie's approach is a bit different though in that he doesn't stress a belief in positivity backed by spirituality. Carnegie speaks more about the manipulation of others through the denial of your personal impulses. For him, people are to be managed and cajoled to come around to your way of thinking. The goal is still to make the sale, but to make the other person feel good about buying your product because they think the world of you. The title of his book says pretty much what it is all about. His principles are many, but all of them are cut from the same cloth. People want to be regarded with respect and consideration for how important they believe themselves to be, so let's take advantage of that for financial gain.

Carnegie maps out some principles for "managing people" in his book. As I read it, I found that some of what he says related to my own principles developed over the years in my work with actors. But the intention behind these principles is where I parted company with Carnegie. The underlining intent to manipulate for financial gain is always present for him. When an actor playing a character considers their partner through objective, they do have something in mind for themselves, but usually the goal is to connect with another character. When they are intent on manipulation, they are making a choice without regard for their partner's truthful responses to what they are doing. They are not really seeing or listening to their partner. They have made their mind up to get the response they wanted, and nothing will influence or change that. This an actor who is unconnected to their partner choosing to self-generate behavior for their own selfish purposes. More on this to come.

For now we can say that a manipulative character is usually also a nefarious character. These typically are not nice people in the context of the play and their actions in relation to other characters make this readily apparent. These are the villains. Even so, the actor playing a villain does well to consider the responses of their partner even if it is only to advance their own nefarious objectives. There is no such determination for Carnegie. Actually, it is quite the contrary. He speaks of these principles of manipulation as virtues that should be sought after and developed by any successful businessperson.

Here are just a few of Carnegie's principles as outlined in the book and my commentary on each of them:

Principle: "Become genuinely interested in other people." (Carnegie 1936)

A nice idea, but the intent is to have this person like you as the title of this chapter suggests, "Six Ways to Make People Like You" (Carnegie 1936). How can you be genuinely interested in a person when your goal is to be liked by that person? It seems more likely that you will feign interest in that person to have them see you in a favorable light. As we will discuss in another chapter, being liked is not an objective. The focus on gaining the approval of another puts you on a road towards a misrepresentation of the relationship. In our work as public speakers we are focused on sharing our research. We do take our partner the audience into consideration for this, but for the purpose of them recognizing the importance of the work we do and not for their approval of who we are. Actors work to find what is interesting in their partner and to respond to that quality, but their intention is not to be liked, unless of course as I've said they are playing a villain.

Principle: "Be a good listener. Encourage others to talk about themselves." (Carnegie 1936)

Very useful advice. We could all be a little humbler and talk less about ourselves, but once again the intent is about something else. Humility is a good quality for anyone to have especially actors who believe that "my partner is the most important person on stage." This is a principle of great importance to the actor. To be humble and to honestly want to know about the lives of others is admirable, but Carnegie is always interested in what can be gained from the other by listening to them. His examples always end with that listener selling that person a car or an insurance policy.

Principle: "Make the other person feel important and do it sincerely." (Carnegie 1936)

Well, the irony is built into this one. If you have a plan to make another person feel important for your own personal gain, it seems doubtful that this may be accomplished with sincerity. The title of this chapter states the true intent, "How to Make People Like You Instantly" (Carnegie 1936).

What Peale and Carnegie and countless others who have written books about success have in common is the idea that your nature is the problem. You are standing in the way of your success. You and your negative thinking have caused you to feel inadequate or inferior. In short, you are to blame for any feelings you have of fear or apprehension about public speaking or anything else for that matter. To be successful, you must simply pull yourself up by your bootstraps, make more positive choices and by the way deny who you are.

There is some truth to these ideas of course and there are many people who will take the advice of Peale and Carnegie and run with it. These people are comfortable in their more extroverted personalities. If you see yourself in this category, you may still find the approach of *this* book to be helpful, but for different reasons. Some of these involve looking beyond yourself and opening your consideration for how others think and feel. The focus is on others but not for the manipulative goal of personal gain.

But what if you are in that other category? What if by nature you are a more thoughtful and understated person? Can you still be an effective communicator? I think you can. In my view, you are who you are. You don't need to apologize for whatever nature has provided for you. This is true for the extrovert, the introvert, and everyone in between. I believe there is a way for you to acknowledge your nature and share what you know about your work with others, whether the thought of that thrills you or fills you with dread. I believe there is a way for you to actively participate in your community and be a contributor to your field no matter who you are.

I don't want to diminish the ideas of either Peale or Carnegie out of hand. They had a particular point of view and a right to share it just as I am doing here. But for me there is an essential difference in the work of an authentic person in the role of communicator from that espoused by these authors and others. For both Peale and Carnegie, the result is the most important thing. Your intuition or your feelings are to be managed in some way and perhaps even conquered for you to be successful. Authentic feelings of doubt or apprehension are holding you back. These things should be ignored. But I don't see it that way. These things are just a natural part of who you are, and you don't have to give up who you are to share your work with others. You don't have to manipulate people to connect with them. Manipulation is the opposite of connection in my view. I hope this book is about connection, but that's just me. You will have to decide for yourself.

CHAPTER 3

Intuition and Its Uses in Communication

Abstract This chapter establishes some of the theory behind the approach to public speaking as I see it. The relationship between creative experience and the scientific method is also discussed here seeking to make connections between the use of intuition and critical thinking.

Keyword Intuitive communication

TAPPING INTO THE INTUITIVE: TRAINING YOUR SENSORY EQUIPMENT

What is Intuition? The ability to understand something immediately without the need for conscious reasoning. This is your brain on autopilot. There is processing of information here that is outside of your awareness. In other words, a hunch. Haven't we all experienced this? That moment when you just know something. You are faced with a situation that calls for a decision and instinct or intuition takes over. Have you had the feeling that it was time to leave a job or a relationship? Did you know why exactly at the time? Perhaps, but probably there was also some bit of feeling involved in your decision. You knew it was the right thing to do in the moment.

This may sound magical, but it's not. It is more common than you may think. You just may not be conscious of it when it happens or perhaps you haven't given it your conscious attention that often. This is not some kind of mystical touchy-feely idea. It is really only half of an idea. Intuition is only the first part of a two-part process. To simply have a hunch about something is not enough and believe me the objections I get from scientists are loud and clear about this. But this is the elegance of the use of intuition. The first step involves that feeling, the knowing without the benefit of reason. The second step

D. Dannenfelser, *The Art of Effective Science Communication*, Palgrave Practical Guides in Communication, https://doi.org/10.1007/978-3-031-57030-8_3

involves making a choice. The feeling is just the beginning. Once you have that moment of intuition you are then prompted to make a choice. What are you going to do about it? Here is where your critical thinking plays a role in completing the action.

The process may be thought of as a marriage between the involuntary and voluntary functions of the brain. The balance between impulsiveness and overthinking. Intuition without critical thinking is incomplete and not very useful from a scientific point of view. To simply feel something is not enough. We are hard-wired to be active, to move forward and act upon our feelings. This may be negative or positive, but it usually follows just the same, human beings act upon their feelings. Sometimes that means suppressing them. Sometimes that means using them to accomplish something by making the feeling tangible in the real world. Either way we seldom ignore our feelings, and this is what makes the use of our intuition so important.

When we train ourselves to think critically about the implications of our hunches, we are on track to getting in touch with our intuition. Harnessing that power to aid in our decision making is a very efficient way to manage our tasks, paying attention to not only what we think, but to what we feel makes for a more complete process for problem solving. When we face questions and challenges by listening to our brains and our guts, we engage a holistic approach to becoming more. Potential energy is just that. The whole self contains thoughts *and* emotions. The effective communicator pays attention to both.

EXPERIENTIAL LEARNING

To go through something is to learn something. I can't tell you how many times I have introduced an exercise or a theater game and have been met with resistance to the playing of the game. This phenomenon is quite prevalent in the scientific community, but I have seen it with so-called acting students as well. The energy that some will put into getting over a game or going around it in some way is amazing.

A simple example for the work-around approach to game process happened at a training session I attended at the Alan Alda Institute for Communicating Science at Stony Brook University on Long Island, NY. I was present for a three-day workshop and sat in on one very interesting event. The Alda Institute is a wonderful resource for the study of Science Communication. People from all different disciplines from around the world participate in these introductory training sessions and some like me return to their home districts to develop classes and workshops of their own.

In this one session, the trainer presented a popular exercise used at the institute and by Alda himself when making personal appearances, some of them filmed for television. So, I am not giving away any secrets when I describe the exercise to you. In this exercise, a volunteer steps up and is given a glass of water. The glass is about half-full when they accept it. The challenge is to cross the room with the glass of water without spilling any of it. Easy right? It is at first, and the volunteer usually accomplishes the task without incident.

But on a second try the idea of stakes is introduced. The volunteer is asked to cross the room again, but this time there is a consequence for any water being spilled. If even a drop is lost from the glass, an entire (fictional) village of people will die. That may cause some nervous energy, but the trainer also fills the glass to the brim this time before sending the volunteer on the journey across the room. This in turn increases the probability of failure for the volunteer who now holds the fate of an entire village in their hands.

The use of the imagination is at play here sufficiently fortified by the introduction of the stakes of a consequence albeit a fictional one. But the reality of the circumstance can be powerful even when imaginary and people do get nervous when they play this game, most people that is. On the day I witnessed a particular volunteer assigned this challenge, he made a choice before walking across the room. He accepted the glass of water but before starting on his journey, he took a long sip from the glass thus lowering the water level and his odds of failure. Clever, yes. Funny, yes. We all had a good laugh in the moment. Why not better your odds if you have the chance? But this guy was forgetting one thing. The point of the game was not to get over on the game. The point of the game was to play the game. He did not accept the challenge, he worked around it. In other words, he cheated.

Some of you are thinking about Star Trek and the Kobayashi Maru right now. If so, you are a true science nerd and I respect you for that, but I will talk about Star Trek in another chapter so stay with me for now on this point; when you concern yourself with being clever, funny, or even interesting, you are in the realm of result orientation. When you cheat and devise a work around, you are avoiding the POC or Point of Concentration of the game. This was not an occasion whereby the player became immersed in the objective of the game. This was a performance. It was a performance for the benefit of the audience, a demonstration of his cleverness, a reassertion of something he already knew about himself. Did he learn anything new at that moment? By side-stepping the challenge of the game, he avoided the POC to play the game and by playing to have a new experience and to learn about something he didn't know. This is a very hard thing to do for some people. For scientists it is often unfathomable to allow yourself to be seen as vulnerable, unknowing. The artist too will sometimes succumb to this manner of thinking. In both cases, the creative instinct is being suppressed for fear of rejection or failure. Let me explain and offer another example.

The Point of Concentration or POC is the objective of the game, the thing we are all trying to accomplish. Each game has a specific POC, but often there is a secondary POC. This secondary POC is simply to play the game. If you can just do that, you will more than likely achieve the original POC. Unfortunately, this can prove to be difficult for some players.

Another example can be found in a story about another game I have played with students in both my acting and science communication classes, the alphabet game. Some of you may be familiar with this challenge, but I will explain for those who are not. All the players sit in a circle. The challenge presented by the game is for each letter of the alphabet to be spoken one letter by one person at a time. If any two people speak at the same time, the whole group must go back and start at the beginning. The POC is to get to the end of the alphabet by having only one person speaking a single letter at a time. Simple right? Well, simple ain't easy and when the game is played as described it can be very challenging and take a bit of time to accomplish the Point of Concentration; one person, one letter at a time.

But the game can take even longer when players in the group decide to try to "manage" the experience in some way. The focus for these people is to get over on the game rather than to play it. It is important to note that no discussion is allowed while playing. This is to avoid any planning of strategies between players. The first sign of avoidance of the POC is the chatter that starts to roll around the circle. There is no connection to the game at this point, only the active avoidance of the playing of it. Some do try to cue others verbally, but this is soon redirected by a simple reminder from the instructor to keep quiet except for the utterance of a letter in succession of the alphabet.

A relative quiet may now be achieved, but still, planning does happen regardless of the no discussion rule, though not in an overt way. Sometimes a spontaneous occurrence unfolds when the players decide as a group to simply go in order around the circle. This is an organic unfolding of a strategy that often happens in the moment. But the goal here, whether the players are aware of it or not, is again to get around the game rather than to play it by going through it. My response to this subversion of the game is to speak up at the same time as one of the players speaking in order around the circle. This of course messes up the plan and sends us all back to the start.

A more deliberate subversion of the POC happens when the players interested in managing or winning the game will soon see that they can organize the other players by establishing a pattern. Some go so far as to point to fellow players as if to signal them as to their turn according to some order they have decided should be followed. It is interesting to see who goes along with this tactic and who refuses to be ruled by these self-appointed leaders. In either case these are the thinkers in the group, the ones who believe they can think their way through any problem or challenge. Little use of intuition is present in these players though they can learn to listen to it just as easily as anyone else. These people pride themselves on brain work and usually have the academic history to prove it.

We have done a good job of promoting such a response in our educational system. There are many rewards for the thinkers, and just as many reprimands and redirections for the feelers. Working with scientists, I have seen evidence of how the brain work approach has been reinforced in them as I observe their behavior in the game. The focus on results and intellectual problem solving in our educational systems is on full display in the alphabet game. It seems we have encouraged our children at a very young age to get the right answer and to present their evidence of it.

This is of course how science works, and it would not be a good idea to have a scientist who only works from feelings. But the balance of the head and the gut may make for a more complete scientific method. The elegant bridge between impulses and over thinking may give rise to a more complete scientist or learner for that matter. We have neglected this part of our development in children. The games we play in pre-school and kindergarten are soon replaced with lessons that focus on us as thinkers rather than feelers. I believe we can be both and be more efficient as both. But I digress. Back to the alphabet game.

When the organizers start to signal to the minions to how they should behave, my response to this is once again to speak up at the same time as one of the minions who is following orders. Back to the start we go, and it soon becomes apparent to all that the only way through the game is to play it. This is where things really get interesting because it is not until everyone has given over to the simple objective of the game that the actual playing of the game begins. Surrendering to a simple task only accomplished through cooperation with others is what this game is all about. There is nothing to win, no prize to obtain, no victory to claim. The playing of the game is the point of the game.

Once everyone allows for that true collaboration an intense silence replaces any other sounds in the group. The players are focused and concentrated on one thing, playing the game. It takes some time for each to feel when they should speak or remain silent, but soon they are doing just that. Long pauses are followed by quick letters in succession. Mistakes are still made, but the focus is changed. Now everyone is working towards a common goal. Now everyone is keeping the POC. Now everyone is engaged in the attempt without any regard for the result orientation. There is only the playing of the game.

When observed from outside the circle it can be seen as very intense. It varies from group to group, but if all are truly playing the game, the end of the alphabet is soon found. It is not unusual for a cheer to spontaneously erupt when this happens. The release of that kind of energy is palpable and felt by everyone in the room. Doing as opposed to avoiding. Total involvement in the POC. This is what experiential learning is all about.

SCIENTIFIC METHOD AND CREATIVE EXPERIENCE

The Scientific Method has been established as a standard for all scientists. This order of procedure has made it possible for the scientific community to determine the efficacy of advances in the field. The steps are simple and clear.

- Make an observation
- Form a question
- Form a hypothesis
- Conduct an experiment
- Analyze the data
- Draw conclusions

A scientist first makes an observation which prompts a question. The question causes the scientist to explore and research the background information needed to form a hypothesis. The hypothesis forms the basis for an experiment to test this new idea. The data generated by the experiment is analyzed to lead the scientist to certain conclusions. These conclusions are then presented to the scientific community for review and process. Challenges may be made to the conclusions or support may be given. Either way, the process is repeated each time a new hypothesis is formed, and the conclusions are submitted for review. This process is closely mirrored by the purposes and goals of the actor in their commitment to the creative experience. Let's look at how the scientific method for the scientist can be compared to the creative experience for the actor more specifically in the following table.

Scientific method	Creative experience
Make an observation: The scientist makes an observation about the world around us.	The actor does the same always looking into the social, cultural, economic, political, and religious structures that form the constructs on which the world of the play revolves. These are represented in the fictional text that is analyzed by the actor through table readings and discussions.
Form a question: The scientist is prompted to form questions about their observations.	The actor too is constantly questioning and trying to uncover the underlying motives or objectives behind a character's behaviors and actions.
Form a hypothesis: The scientist forms a hypothesis based on their observations and questions.	The actor forms a hypothesis for the truth of a character they portray by determining their motives and forming a world view for that character that justifies their behaviors and actions.
Conduct an experiment: The scientist conducts an experiment to support their hypothesis.	The actor forms objectives that are tested in their relationship to fellow actors and their portrayals of other characters. The rehearsal process is their experiment and the stage space their laboratory.
Analyze the data: The scientist analyzes the data that is generated by their experiments.	The actor takes in and responds to their fellow actors receiving immediate validation or rejection of their objective choices.
Draw conclusions: The scientist draws conclusions from their experiments and presents these to the scientific community for review and criticism.	The actor discovers the most effective dramatic choice of action needed to fulfill their objectives and adjusts those that do not succeed. The actor then performs for the public on a stage and lets the audience decide whether their portrayal of a character rings true or false.

The scientific method and the actor's process of creative experience are not so dissimilar when you make the comparison. Both individuals have an approach to the work they do. Both must consider limitations and careful analysis to proceed with purpose. Both are engaged in a critical and creative process and ultimately must answer for the results of their work. The scientist is held responsible and accountable by the scientific community. The actor by an audience in attendance at a performance.

The critical and the creative are not at opposite ends of a spectrum. They are rather more like two sides of a single coin. The scientist who doesn't consider themselves to be creative is missing an essential part of the scientific method. To make observations and form questions are inherently creative activities. To further develop these into a hypothesis that will ultimately be shared with others is performative in the best sense of the word. Like the actor, the scientist must make use of intuition and imagination to create something new.

CHAPTER 4

Acting Techniques Applied to Science Communication

Abstract This chapter is as the title suggests. Here the application of acting techniques to the development of skills as a science communicator is outlined and explained. Concepts such as generosity of spirit, actor's faith, and telling the truth are discussed to lay a foundation for understanding how the work of an actor can be studied and applied to the goals of effective communication.

Keyword Acting techniques

ANOTHER WARNING

Yep, I'm going to be talking about actors again. Don't worry, I haven't forgotten everyone else. It's just important in my view to understand what these artists do to make sense of how it may have implications for the rest of the world. Please stick with it. We'll get back to you soon.

ACTING IS DOING

Ask a person without much experience of the craft of acting to define what it is and usually you will hear something about emotion. An actor emotes or shows emotion or is an emotional being who has little trouble crying on cue. Not so. An actor is concerned primarily with what can be played. Emotion cannot be played. Attempts to show emotion for an actor amount to what is known as an indication. I am sad and so that is why I have this frown upon my face. This is what sadness looks like. What anything looks like is immaterial to the actor. That is a result orientation, and it holds little interest for the serious actor.

Actors are doers. The actor's focus is on the verb rather than the noun. To be sad is a state of being, a noun and therefore in-actable. Conversely to do

D. Dannenfelser, *The Art of Effective Science Communication*,
Palgrave Practical Guides in Communication,
https://doi.org/10.1007/978-3-031-57030-8_4

something, to pursue some action, a verb that results in a feeling of sadness is much more active and so more actable. An actor experiences feelings because of some action. An actor *has* their feelings, but they *play* their actions.

When an actor commits to an action, they are encouraging involvement in themselves. They are choosing action as a means towards involving themselves in a conflict as described by the playwright or screenwriter. They are seeking to honor the intentions of the author by doing. This starts with an idea or an objective that is then physicalized by the actor through doing. The actor does something in response to a cue from the circumstance as provided by the author and transformed into behavior by their fellow actors.

Feelings are never a part of the discussion at the table when actors go about their work in analyzing a text except to say how a specific objective has conjured them up. Wants and desires are very much discussed and will consequently prompt the actor to choose an objective that will fulfill that want. More specifics will be covered on objectives when we talk about text analysis related to speeches in another chapter. Suffice to say, professional actors are doers and will always be doing as they practice their craft. Crying not prompted by a specific want or desire is for amateurs and babies. I take that back. Babies always cry for a reason.

First Principles

There are a few principles that actors follow to set the stage for their work. It should be said that different acting programs have different principles or schools of thought on the work of an actor. This is as it should be. After all, acting is a craft and an art form. In other words, it ain't math. Sometimes 2 + 2 in the acting world can equal 7. And it is fair to say that there are as many different points of view on acting as there are actors, but one thing is certain; when you find the one that works for you use it.

Philosophical discussions aside, for the purposes of our work we believe in the approach to acting as developed by the Russian actor, director, and teacher Constantine Stanislavski and his American interpreter of the art form, Sandford Meisner. These two theorists did much to shape the world of modern acting and have proven to be very useful to us as we try to apply principles of the field to those required of the student of public speaking. What is an actor if not a communicator?

There are basically three questions that form the basis for understanding the work of an actor. Once answered and understood these become the principles that actors live by, at least those actors who find themselves enrolled in the conservatory training program at Rutgers University's Mason Gross School of the Arts. I have worked in both the conservatory and the BA theatre program at Rutgers and am well acquainted with these questions and others. Let us consider them now.

The first of these is concerned with the students' understanding of a working definition of acting. What is the *definition* of acting? Well according to our approach; acting is living (doing) truthfully under imaginary circumstances. As I have already stated, actors are doers. It stands to reason then that any definition of acting would have to include that idea as a part of it. Living or doing truthfully under imaginary circumstances may be hard to fathom for the non-actor, but hopefully our elaboration on this idea in this book will make this concept more understandable.

What is the *fundamental* of acting? This is the second question. The fundamental of acting is that my partner is the most important person on stage. That's right. The most important person on stage is not you. This may seem well, fundamental, but it may be surprising to the neophyte. This may be perhaps because of another common misconception about actors besides the whole emotion thing. It is often thought that actors are vain and self-involved creatures. While that may be true of some celebrities, the serious practitioner of the craft of acting understands that without my partner, I am nothing.

Try telling that to an auditorium full of young first year medical students (I have). Sorry, but this fundamental idea is true. And not only that, it is also necessary and useful to anyone who wishes to be a successful communicator. My audience (partner) not me is the reason I am here. Those same medical students are challenged to answer the question; if there isn't a patient (partner) what makes you a doctor? Practically speaking, you don't enter an empty examination room to do your doctoring. You enter because there is a patient there waiting to be seen and heard by someone who may be able to help them or at least allay their fears. Without a patient you have no reason to be a doctor.

The actor understands this concept and more. Actors are alive to everyone and everything around them. The actor understands that my reason for being here is ultimately connected to another. Without a partner I have nothing to take in and respond to in the staged reality. The actor's work is guided by the idea that taking-in and responding is the way of acting. And so, I take in everything in the imagined space, and I respond to it, most importantly my partner.

In this sense a partner can be defined in many ways. I can view my surroundings as a partner. The set on which I am working, which includes every sight and sound I experience, should become alive to me as I take in the room in which I find myself. The lighting I see before me that illuminates the space is my partner. The sounds and or music I hear as a part of that story we are telling are my partners. And of course, my fellow actors are my living, breathing partners in the staged reality. Together in partnership we embody the intentions of the playwright. Serious actors understand this idea of dying to self. Communicators of any kind should as well.

Question number three. What is the *principle* of acting? If in fact I believe that my partner is the most important person on stage, then I will be right in line to practice and test this idea through what we call the *principle* of acting; I don't do anything unless my partner does something that causes me to do it. That is the principle of acting in line with its fundamental. I know this is a kind of a chicken and egg thing, but somehow, one person in the relationship picks up on something happening in the other either consciously or unconsciously and we are off and running, taking-in and responding through the relationship in the imagined reality.

The great American actor and teacher Sanford Meisner referred to this principal as the "Pinch and the Ouch" (Meisner 1987). His contention was that if you don't get pinched you don't say ouch. This of course made for some colorful first meetings of his classes in New York City. He would wander around the room while speaking and then suddenly grab a hunk of flesh out of the back of some unsuspecting student and twist. The resulting shriek from the student became exhibit A as Meisner would quickly exclaim, acting! Of course, today he would most likely be brought up on charges of abuse. This was a different time. We can make the case today without the use of his physical example and hopefully still have the impact he intended. An actor is a doer, yes, but the actor only does in response to a partner and every subsequent choice of doing is guided by that pinch/ouch sequence.

This is another one of those instances where the ego comes into play. Actors or any communicators who determine to put themselves at the center of the experience will soon fail to capture the interest of an audience. Actors are hired to create behavior on stage. Behavior, when it isn't motivated by any real or imagined stimulus, will always seem false or inauthentic. Conversely, a true response to a partner in the moment is an exercise in authenticity even if the circumstance is a fiction dreamed up by an author or by you for that matter in the form of a speech you are sharing.

When you place yourself at the center of the experience the audience becomes witness to something different. This is what we call self-generating behavior. The focus here is on the individual working within a bubble. In this bubble nothing matters except for the individual's own selfish impulses. They don't act in response to anyone but themselves. It isn't only indulgent; it's exceedingly boring to watch. Ironically, this kind of self-indulgence can be motivated by fear or feelings of inferiority. Just like that guy who sipped from the glass of water to avoid playing the game, the self-generating actor or speaker believes they must be interesting, clever, or funny to hold the attention of the audience. They are afraid that if they don't generate something in the moment that the audience will lose interest. This is a demonstration of a lack of understanding of the process. This person is not rooted in any clear approach to the task at hand. This person is operating under the assumption that "I am responsible for everything that happens in this moment. I have the entirety of the responsibility for what happens here today."

The ordinary healthy and normal person who believes in this way of thinking will suffer a great deal of anxiety when faced with the task of acting or speaking publicly. The unhealthy narcissist will blame the audience and refuse to see how they themselves have contributed to the failure of the moment. I am assuming I am talking to the normal folks out there when I ask, why would you do this to yourself? You are not the universe. You are human and as such you have that in common with your audience. You need them to complete the experience.

Humility is the key to every great performance. The actor who understands this will work hard to raise their partner up because they know it will always come back to them in the end. I alone am not responsible for the success of our experience today. I am counting on you to take in what I have to offer and to respond. I will in return continue that pinch/ouch sequence with you and together we will both create the experience we want to have. Our talk or performance as it may should feel like a conversation with the audience (partner). The healthy normal person will understand this. The narcissist has already thrown this book across the room. Imagine their frustration if they purchased this as an eBook.

FOCUS AND PRESENCE, REALLY SEEING, REALLY HEARING

To really see and really hear your partner is to be present in the moment. Let me return to the doctor/patient relationship for a moment as an example of this concept. It may seem obvious to say, but the fact is a doctor cannot be a doctor if there is no patient. Yes, I know there is research, but for whom is this research being conducted? If we consider the face-to-face doctor/patient relationship, it is accurate to say that the only reason for a doctor to enter an examination room is to tend to the patient who waits there.

This is an obvious statement, right? But the looks I have seen on the faces of first year medical students when I say this are precious. For many of them the look is one of recognition and acknowledgement, but for some it is surprise. Believe it or not, some of them have never considered this idea. The relationship between themselves and the patient is not necessarily central to their view of their work as a physician.

What? You may say. How can this be true? Well, you need only look back at the concepts we have been exploring to see how one's own anxieties or insecurities can often lead one to put themselves at the center of every experience. I'm the one putting in all this time and effort to become a doctor. I don't recall seeing any patients in my room when I was studying for those exams. I'm the one who knows this stuff. I took it seriously and worked hard to get where I am.

And where are you exactly? In a relationship with a person who has come to you for help. Your knowledge only has meaning in the context of how it can be of service to another. Once again, narcissists beware. I'm talking about thinking of others, taking in and responding to others. If you've retrieved this book from the corner where you threw it, you might want to pitch it again. I'm sorry, but without a patient, you are not a doctor.

Of course, the opposite is also true. That person is not seen as a patient outside the context of their visit to you. They are in fact only thought of as a patient when they are in the circumstance of visiting you, their doctor. It works both ways and that is important. By each of you playing your part, you become something specific to each other. The relationship is in service to a specific desire that exists between the both of you. The patient has a desire to be well. The doctor, who isn't a narcissist, has a desire to make them so.

How then does this apply to the scientist or to any public speaker for that matter? I'm glad you asked. What is a public speaker without a public? This is a person alone in a room talking to themselves. Fine for rehearsal, but this is not the real show. That conversation that happens between a speaker and an audience is not unlike what happens between a doctor and a patient. Together we have a common goal, and we need each other to get to it. I have something to say to you today, but I need you to also hear it and to let me know you have heard it by giving me a response. This is not always obvious, a person from the audience raising a hand and asking a question for example. Sometimes it happens in the form of sounds from the audience, coughing, rustling as they shift in their seats or if you are doing your job effectively absolute silence. No sound at all is sometimes a very good thing. It means they are listening.

To go back to an earlier idea, this is where you must feel what is happening in the room. This is the intuition at work. That pin dropping moment is gold to the public speaker, but once again it is a result of something you are actively doing and so it is not your focus or intent. You are busy pursuing an objective in the moment and have no control over what is happening out there. Even so, you can feel when what you are doing is working and the conversation has begun. At this moment you must do your part and keep that energy alive by doing what you came to do, speaking specifically and directly about a central idea and really seeing and really hearing the audience as they respond in the moment.

PLAYING FOR KEEPS AND REALLY DOING

We have covered several concepts so far, but none of them will lead to a successful experience either of acting or public speaking if you don't commit to actually doing these things. Whatever your apprehensions may be, you must decide to really *do* in the moment. Acting is about doing remember and so your commitment to action must be 100%. Following the principles and focusing on your objective will help, but if you haven't decided to *do* these things your efforts will be for naught.

Playing for keeps means you have given yourself permission to be present and to share your work with others. You have refrained from result-orientation and have refused to judge yourself. You have agreed to forgive yourself when ultimately you succumb to self-judgment and you are busy going after what you came here to do, sharing your very specific, central idea with an audience. You've also honestly put in the time to rehearse and prepare for this moment and are ready to leave that homework at home so you can be present and in the moment. When you commit 100% and play for keeps, something wonderful happens, you forget to be anxious. You are too busy to be afraid. You have a role to play and work to do.

SAYING YES AND SURRENDERING TO YOUR ACTOR'S FAITH

To be an actor is to say yes to things that scare you, make you uncomfortable, or challenge some belief you have about the world around you. But an actor must have faith in their own ability to survive by trusting in the process. To say yes is to pursue a positive, to always be moving forward. So why do would-be communicators and even actors sometimes say no when faced with a difficult challenge as a character in a play or as a speaker sharing a speech? The answer is ego in my opinion. I am at the center of this experience and my anxiety about it is determining how this will go. This again is a function of the kind of self-involvement that leads to the self-generating behavior that makes for some ineffective acting.

The actor and the speaker are challenged to have faith. But saying yes is not just about blindly accepting a positive point of view. This is a conscious choice to believe in the work you are doing and to believe that it has merit and value for others with whom you make contact and contact is ultimately what all this effort is about. Saying yes is not about pushing forward some agenda you have. It is about recognizing all the agendas in the room and reconciling them all as acceptable. To do this you must first take stock of where you are in your own process and then proceed from there. You don't push, you allow.

Some years ago, an actor friend of mine told me a story of a difficult time he was having with a monologue in a play he was rehearsing. The director was asking him for something very particular while at the same time being very vague about what that was. Without a specific path for him to follow he tried to generate a lot of interesting, but non-specific qualities for the speech. All of this was being pursued from a generalized perspective and a great deal of effort on his part to make something happen. This became a very result-oriented process that belied all his training. Actors can do this too, you know. They can forget that the action is the key to unlocking every choice.

After countless attempts to find a quality rather than to pursue an action, my friend gave up and just sat down on the stage to brood. When the director asked what the problem was, he simply said, "I'm tired. I can't do this anymore." And just like that an idea seemed to come to the director like the flipping of a switch. "Do it from there," he said. "Let yourself feel your exhaustion and do the speech from there." That was it. My friend had been sending out so much energy trying to make something happen when all he needed to do was surrender and have faith in the moment. He was tired and denying the truth of what he was feeling which was keeping him from finding the path to the truth of the monologue. His own lack of faith was getting in the way. He also returned to his process and made a choice about a specific objective for the moment, but the quality he was looking for came as a result of him looking at where he was in the moment and having faith in what he was feeling.

When we are anxious about speaking publicly it is often a symptom of a self-generating idea of who we *should* be. Coming from where you *are* is important. Recognizing your anxiety and not denying it gives you ownership of it. You don't have to fight it. You accept it and then come from that place of fear as the reality of the circumstance. When you own it that way you begin to understand that you have a way of proceeding. It is simply a fact that you are scared, but that doesn't change all the work you have done to prepare for this moment. You say to yourself, "This scares the hell out of me. Here we go."

By the way, I had some of the same feelings about writing this book. It scared the hell out of me. But I decided to believe I had a right to share my voice on this subject and do it anyway. Whatever happens next is out of my hands. I suppose I will just have to have faith in my own process.

GENEROSITY OF SPIRIT

What is meant by GOS or generosity of spirit? GOS grows out of a desire to make contact and it can be a very powerful force in our work and our lives. As human beings we seek community as a means towards our survival. But the search need not be that difficult. We need not send out a lot of energy to generate a feeling of community. Being open to others and taking in what is happening in our relationships can go a long way towards developing that sense of community.

In this way we are sharing ourselves with others by recognizing who they are and giving our focus to them. Our generosity or willingness to really see and really hear another person creates a positive sense of spirit in that relationship. Listening is courteous yes, but it also prompts the same kind of courtesy in others. The give and take that happens on stage between actors comes out of this notion that as my partner you are the most important person here. Because your partner takes the same point of view a spirit of generosity is generated between all concerned and adds a lot of energy to the moment.

Try to be generous with yourself in the way you forgive yourself for mistakes. Try to share that same generosity as you forgive the mistakes of others. It is easier to accept your limitations as a public speaker when you recognize that the audience is made up of flawed human beings just like you. Actors and Public Speakers do themselves a favor by opening up to others. When all parties are engaged in this kind of positive spirit, many difficult tasks are transformed into challenges that can be accepted and even relished in the moment. "This scares me, but I'm not doing it alone. We are in this together. Here we go."

TELLING THE TRUTH

These concepts also come to life because of a simple idea, what I have to say today is true. The actor who is speaking from a text that was imagined into life by a writer has found the truth of those words through the rehearsal process. "The reality of the fiction," as my teacher Paul Austin used to say, is about finding the universal truth in a text. The truth of the story is not the point. The truth that is told through the story is what you are after as an actor. That is a deeper universal truth that is rooted in a common understanding of a shared experience.

How well the audience relates to what your central idea is depends upon their own common experience of it. Knowing your audience will help you to speak directly to them. When you know who they are, you also know what concerns them and how those concerns are shared by you. You can then find the language that will connect you to them as everyone works towards a true understanding of the issue at hand.

The best advice I've heard on this subject came to me when I was working as a playwriting instructor in a program at the Public Theater, the home of the New York Shakespeare Festival in NYC. I was sitting in on a rehearsal of *The Winter's Tale*. The actor Mandy Patinkin was a member of the cast and he graciously offered to speak to the small audience of educators present that day. When asked by someone about his thoughts on acting he said simply and directly, "plant your feet, look your partner in the eye and tell the truth." Good advice for would-be actors and public speakers alike.

CHAPTER 5

The Development of Ensemble

Abstract Taking note of a particular aspect of the actor's process and identifying it as a key to the development of any productive relationship, ensemble building is considered as an effective means towards achieving the goals of the communicator seeking to connect to their audience. This chapter makes the case for teamwork and the view of the audience as partners and fellow ensemble members in pursuit of common goals.

Keywords Ensemble building • Practical games • Exercises

What is ensemble? It may be said that an ensemble is formed when any gathering of like-minded people occurs for the purpose of participation in some process or action. Ensemble is a gathering, but it also is a function of a process that precipitates learning. I believe that it is important to form such an ensemble when a group engages in training for the purpose of public speaking.

Theater games have long been used as a means towards the development of ensemble. The game form is interactive and encourages collaboration between players. The group agreement that is necessary for the playing of a theater game is also a part of the ensemble building process. Each of these things, when pursued actively, can lead to a feeling of connection and ultimately a better practice of communication between the individuals in the group.

If connection is our desire and communication our goal, then the building of ensemble becomes a significant step in that direction. Each time you share your speech with an audience you are joining together with them in a group agreement with a common interest and goal in mind. This is ensemble.

How Do We Form an Ensemble?

From an improvisational point of view, play is at the center of every effective attempt to form an ensemble. Whenever I begin a new class, I am faced with a certain challenge. The room is full of individuals, some with a genuine interest in the work and a willingness to take a risk to develop new skills and others who are skeptical of all of it. Honestly, I tend to be on the skeptical side myself. I am not naturally positive in my outlook about new ventures. I am cautious by nature and introverted. I have seen too many ill-fated group meetings lead by well-meaning, but untrained individuals who have tried to gain my trust without earning it. If I ask you to place your trust in me, I must first prove that I am worthy of it.

As I have mentioned earlier, trust is at the center of all the concepts I have shared with you. My willingness to take the same risks as those I ask of a student is essential to the success of my efforts to gain trust and ultimately to form a strong and functional group, an ensemble. If I ask anyone in the group to participate in an activity, I must be willing to participate in that activity myself. Of course, it helps if the objective is an honest one that keeps my partner at the center of my consideration. I can't pursue an objective that is manipulative and presupposes a relationship that puts me in a position of power over my partner. Once again, the assumption of goodwill is my guide, and my generosity of spirit must be in full play as we seek to form an ensemble together.

The Nature of the Games

Play is the equalizer, and the game is the form that provides the structure for our play. It is the common goal found in the game that binds us and strips away the roles of student and teacher. Playing games together is a process of connecting us together. These games must be collaborative in their structure. All players involved in the game must need each other to complete it.

Theater games are purposely structured in this way. As I have mentioned earlier, every game has a POC or point of concentration. This is the central agreed upon goal of the game, the thing everyone is striving to achieve with the help of others. Until the point of concentration is discovered and agreed upon by the group the game may seem to be incomplete. The games are like puzzles. The pieces are all there, but they need to be formed into a whole by the group. You and your fellow ensemble members work together to complete the puzzle. It can be challenging and fun, but it also allows for the development of skills in communication.

To rely on another implies that you must connect and communicate to be successful. It is deceiving to think that developing real skills in communication could be so much fun, but that is the point. The simplicity of play and the structure of the game is a natural and ancient process. One way we have found to survive as a species has been through our collective cooperation. Another way has been through competition. More on that in a moment. We have discovered our own capabilities by "playing" certain roles, trying them on for size and seeing what fits, but we don't do these things alone. Our playing is always in relationship to others. Relationships are at the center of our play.

Theater Games and Relationship to an Audience

It is not an exaggeration to say that a group can meet for a year and play theater games that require no language whatsoever. It is ironic to think that a book on public speaking would emphasize training in a game form that requires no speech at all. The non-verbal games that improv students learn are essential to the development of the language-driven public speaking they will do in the future.

To really see your partner is the first step in becoming an effective writer and speaker. If I am seeing another when forming my ideas for my speech, then I am considering my audience and being sure to address their concerns based on what I see. I can tell from an actual observation of my partner's reactions to me non-verbally what they do or don't understand about what I am doing. Later when I am sharing a speech with an audience, I can see how that speech is being received in the moment and adjust my focus to be understood. Seeing your audience is important because they will also be actively communicating with you the speaker, they are just doing it in a non-verbal way.

The skills involved in being present for an audience start with being present for my partner in the structure of the game. Taking in and responding doesn't necessarily mean I am speaking. Sending out energy without taking in what is happening between you and another is self-generating if you remember our acting terms. To this end, the non-verbal game gives the student a lot of practice in taking in and responding. The same principle can be applied to the sharing of language and more specifically a speech later. If you are truly connecting to your audience, you will be seeing them and taking in their responses to what you are saying and that will change the way in which you speak.

Group Agreement, Contest vs. Competition

The essential element of all improvisational experience is the group relationship. Our group agreement is how we accomplish the challenges put forth in each theater game we play and ultimately lead to the growth and development of our interpersonal skills. As I've mentioned, the games are constructed purposefully to be played cooperatively. The group members of the ensemble need each other to complete the game. This is an organic process whereby problem solving is done collectively to lift the skills and consciousness of all who play. There are no winners or losers in the improvisational process. We all strive together to meet a challenge. If we fall short of the completion of a task, we make a collective adjustment and try again.

Result orientation puts an end to this organic process. When we seek to gain something other than the satisfaction of working collectively to play the game, then we have created an atmosphere for competition. This is contradictory to the game process. Our focus in playing is on contest rather than competition. Together we are engaged in a contest that challenges us, but not towards some endgame of some prize that is an emblem of success. The point of the playing is the playing. Being alive to the process is the reward if there are any at all.

This is a difficult concept to accept for the very competitive people who make up much of our society. We are taught when we are very young to compete. Competition is survival. We have been hard-wired to survive as human beings and for good reason. We evolved into a world that was fraught with danger. It makes sense that we should be on guard. Experience has taught us to beware of new experiences and people. Our instinct tells us to look for threats and to manage these to survive. That often means we look at the world in terms of winners and losers. To win is to survive.

So, it's difficult to consciously put that instinct aside. Support for a competitive point of view is everywhere, in sports, in business, and in our interpersonal relationships. But to become a member of a group, an ensemble, it is necessary to make a different choice and to die to self. To agree to the concept of ensemble, we must see ourselves as a part of something bigger than just me and to trust others who are trying to do the same. The students who try to lead an exercise or get over it while avoiding the POC are in this category, the competitive survivors, but really most of us feel this way at one time or another. It is human nature to look for the patterns in groups that break down into leaders and followers and even more so to want to be the former and not the latter.

Viola Spolin puts it this way, "A healthy group relationship demands a number of individuals working interdependently to complete a given project with full individual participation and personal contribution. If one person dominates, the other members have little growth or pleasure in the activity; a true group relationship does not exist" (Spolin 1963). What she is talking about here is contest over competition. These are individuals yes, but each one puts forth an effort that is in service to the desires of the group.

As a communicator, you do yourself a favor by recognizing that you are a part of something more important than yourself. The anxiety you feel about public speaking is generated by your belief that you and only you are responsible for what happens when you share your speech with an audience. But this is an egocentric point of view. When you think of this experience in that way, you are ignoring the audience and the role they must play as you try to collectively accomplish this task. When I let go of my ego, I am free to see myself and the role I play in this event. I am also free to recognize the roles of others as we create this experience together. It isn't just about me. This is how we all have an experience of ensemble and maybe a sense of success in the doing.

PRACTICAL GAMES & EXERCISES

The Trust Line

Setup and Instructions
- Players stand at the opposite end of the room from the instructor.
- Each player takes a turn to close their eyes and walk forward towards the instructor.
- Players should not open their eyes or stop walking until the instructor stops them by gently touching their shoulders.
- Each player takes a turn.
- After all have taken a turn, the instructor asks that someone from the group stop them as they also walk forward towards the group with eyes closed.

Essence of Discussion
Discuss the threshold of trust—everyone has his or her own threshold of trust represented by the imaginary line on the floor they hesitated to cross before they felt uncomfortable or frightened.

Clap for Action (Austin, Boyd 1973)

Setup and Instructions
- A single player leaves the room.
- All decide upon a physical activity or two they want the single player to do.
- The single player re-enters and begins doing.
- All others respond with applause when the single player gets close to desired activity.
- All are silent when the single player is moving in a different direction than desired activity.

Essence of Discussion
Always be doing is the mantra here.

The Alphabet Game

Setup and Instructions
- Players sit in a circle.
- Each player speaks a letter of the alphabet in the correct order one player, one letter at a time.

- No order for speaking is established.
- If any players speak at the same time the group starts again at the beginning of the alphabet.
- No discussion or preplanning is allowed.

Essence of Discussion
Everyone is encouraged to play this game without any discussion or preplanning. The discussion is centered on one's commitment to keeping the POC and playing the game. How can you be present and focused on your partner by simply playing the game. What kind of work arounds emerge and why? Process rather than result orientation is the goal. The simplicity of really doing is at the heart of every game we play.

The Mirror

Setup and Instructions
- The group is divided up into As and Bs and partners are established.
- As start as the initiator of movement, Bs follow.
- All are side coached to keep the mirror and to move simultaneously.
- After a first round, Bs initiate and As follow.
- For a third round, the group is side coached to follow the follower. There is no leader and no follower, but a fluid motion is developed between each player.

Essence of Discussion
Focus and concentration require you to be present in the moment. Success in this exercise is dependent upon collaboration rather than competition. The challenge is to stay together in one movement. Consideration for the other person is a gift you give to yourself. How can I shift my focus away from myself and take in my partner?

The Group Mirror

Setup and Instructions
- Whole room is a mirror.
- Leader calls out shift and all turn to see another and mirror whatever that person is doing.

Essence of Discussion
An extension on the work by partners on the mirror. Once again, focus and concentration require you to be present in the moment. Success in this exercise is dependent upon collaboration rather than competition but now the challenge is to stay together in one movement that is extended out to include the whole group.

Who Is Standing/Who Is Sitting? Circle

Setup and Instructions
- All players sit in chairs in a circle.
- Leader asks four players to stand.
- Leader calls begin.
- Without discussion, players in chairs change positions.
- Four players stand in front of chairs at all times in the circle.

Essence of Discussion
This game is concerned with connection through observation. To play effectively you must pay attention to the group. Really seeing the whole by paying
attention to individuals brings the players together in the pursuit of a common
goal, to keep balance around the circle.

Ocean Wave

Setup and Instructions
- All in a circle in chairs except for one player standing center.
- Player center calls shift right or left.
- All players in chairs move to right or left.
- Player at center tries to get to an open chair.
- New player at center calls the next shift and the process repeats.

Essence of Discussion
Again, connection is important. You must remain in contact with your fellow
players, but the pursuit of an objective is also explored here in a friendly and
non-competitive way. I say non-competitive, but it can be remarkable how that
instinct kicks in when one player is in pursuit of a chair in the circle while others
are doing the same. My adjustment to the group is to remind them, "it's just a
chair." The real purpose is to communicate simply and specifically through the
action of securing or protecting a chair in the circle.

Chair Game

Setup and Instructions
- All in a circle in chairs except for one player standing center.
- Player center makes a statement.
- Players move to another chair if the statement pertains to them. They
  cannot simply stand and sit again. They must move to a different chair.
- Player at center tries to get to an open chair.
- Two players share a statement in the middle one word at a time (Extension)

Essence of Discussion

Opportunities for workarounds are often observed while playing this game. Some players will not move at all while others will try to stand and sit in the same chair. But this game also provides opportunities for risk taking. If an obvious statement is made such as "anyone wearing blue," players will usually move if they are in fact wearing blue, but if the statement is more personal such as, "anyone who has ever lost a friend," players must decide how much to reveal about themselves personally. When building ensemble, risk taking is essential to form the bonds of connection earned through trust.

PART II

Practice

CHAPTER 6

Writing Your Speech

Abstract Moving from theory to practice, this chapter introduces the approach to speech writing and the point of view taken for the choices that may lead to a more effective speech. Establishing a central idea and making use of literary tools such as metaphor, simile, and proverb to create an accessible and authentic speech are subjects covered in this chapter. A study of dramatic structure is also included here as it pertains to storytelling and the development of a premise.

Keyword Speech writing

THE CENTRAL IDEA

What is the central idea (CI) of your speech? This is the overall point you wish to make. Try to develop this into a concise, positive statement. This is a speech about *what?*

Many times, I will ask a student to tell me what their speech is about. The answer is too often a long explanation about the content of the speech. They cite many examples and give expansive details to justify the content and to justify their sharing of that content with the audience. This again comes from insecurity and self-judgment about the worthiness of their speech and subsequently themselves. When we remove this approval/disapproval concept from the equation we start to get down to the real business at hand.

Your speech must be about something, one simple, central idea from which all other ideas emanate. This becomes an organizing principle for the whole speech and helps to keep the power of the idea at the center of it and not the power of judgment of the person sharing the idea. When you know what that

D. Dannenfelser, *The Art of Effective Science Communication*, Palgrave Practical Guides in Communication, https://doi.org/10.1007/978-3-031-57030-8_6

central idea is and you can state it simply and directly, you are on your way to crafting a compelling and specific speech and ultimately on your way to crafting a compelling and specific performance of that speech. When actors work on a play, they begin by identifying the super objective. What is the overall thing that is desired by everyone of the characters? This becomes the spine or through line for the entire play. It also leads an actor and a speaker alike to an authentic and truthful performance.

The truth is at the center of everything that happens in Hamlet. Hamlet's desire to seek the truth sets up the action in the play. Hamlet has mistakenly been described as passive. He waits for five acts to finally take his revenge upon Claudius his uncle, murderer, and usurper of his father's crown. But Hamlet is not merely waiting, he is gathering evidence. He is determined to find the truth and be justified is his revenge. Without proof, any action to kill his uncle would turn Hamlet into a murderer as well. He is better than that, a more just and compassionate person. Hamlet isn't passive. He is busy. He seeks the truth, and this is the super objective or spine for the entire play. To seek is active and activating because every other character has a response to this action either negatively or positively. This is a very strong organizing objective, and your central idea works in much the same way.

THE CENTRAL IDEA AS A QUESTION

It may also be useful to think of a question that is generated by the central idea. The question becomes a cue, and the speech is the answer to the cue. The notion that someone might want to know what your work is about should be a part of every speech you give. When you are clear with yourself about your central idea, see if you can turn it around and put it in the form of a question. "We must address climate change to insure our own survival." This is a simple, but bold statement about a central idea. The question generated by this statement might be, why should we care about climate change?

To frame your speech as if you are answering a question supposes that someone other than yourself has a stake in your response. The fundamental of acting comes into play once again. My partner is the most important person on stage. My focus is to speak to the concerns of my audience and not merely to speak. When my central idea is a response to a question, the audience is a part of the speech. As my partner in addressing climate change, we are exploring the issue together. Once again, the focus on self is secondary in this form.

To share information about your work in response to a question supports a collaborative spirit rather than an oppositional one. The audience is working in collaboration with you and not sitting in judgment. It also creates perspective for you to be addressing someone else's views and incorporating those views into a relevant response. To see something through the eyes of another brings up many insights into the subject that you may have difficulty finding when all the ideas are generated by you.

The Power of Analogy, Metaphor, and Simile

An analogy is a comparison in which an idea or a thing is compared to another thing to better understand the original. It aims to explain that idea or thing by comparing it to something that is familiar, but different. An example of this may be an explanation of how an x-ray machine works by describing the process of holding an egg up to a candle to see what is inside. Candling was an old trick discovered by farmers to determine if their hens' eggs were fertilized. Many of our present-day inventions grew out of practices from a much earlier time.

Metaphors and similes are tools used to draw an analogy or comparison. Therefore, analogy is more extensive and elaborate than the metaphor or simile you use to describe it. The use of a metaphor creates an image in the listener's mind that cuts through a more technical description of a concept or idea. To say that trees are the lungs of the planet paints a vivid picture in the listener's mind of the process of photosynthesis without getting too technical.

Simile is also useful when attempting to explain a complicated subject to a general audience. For example, to say the valves in the heart work like the swinging doors to a restaurant's kitchen is a useful image that gives the listener an instant understanding of the concept you are trying to explore, how heart valves direct the flow of blood. Like the way the wait staff moves through the kitchen doors in a restaurant to avoid collisions, the valves of the heart work to direct blood flow throughout the body, one way in and one way out.

These two literary devices, metaphor and simile, are tools the effective communicator can use to connect to a lay audience. Metaphor is more direct using *is* or *are* to describe, while simile uses *like* or *as* to make a more casual comparison. These devices are at your disposal when you are trying to explain your research which can be complicated and technical. Metaphors and similes can get to the heart or essence of the meaning of your research and therefore make it more accessible to the non-scientific mind.

Proverbs

Proverbs are also literary tools which incorporate metaphorical images to express an idea.

A proverb can be useful when trying to get to the heart of a scientific or technical concept. To follow up on our example of the farmers and their chickens, "birds of a feather flock together" is a common proverb. The more prosaic way to describe this idea is to explain that people of common interests and backgrounds will often seek each other out and congregate together to affirm their shared beliefs or goals. Both are true, but the proverb gets to the essence of the truth in a more economical and interesting way.

FORSAKING YOUR SCIENCE

The simplest way to communicate a challenging concept or term is to just say it another way. Serious scientists will sometimes balk at the idea of "dumbing down" their science for a general audience. But remember the goal of effective communication is to be understood. The use of metaphors, similes, and proverbs to speak about your research does not mean you are throwing that research away. You don't have to forsake your science to make it more accessible. Using this technique, you can speak in very complicated and technical terms and then simply cite an example of those terms in more commonly used language expressed through metaphor, simile, or proverb.

When trying to communicate difficult or overly technical information it may be helpful to make use of these literary tools. When writing your speech, a pedantic description of your research may be accurate, but what does that matter if you have lost your audience in the process? Once you know what your central idea is, start to explore ways of expressing it by drawing analogies using metaphors, similes, and proverbs. If you can do that, you may find that you have already incorporated another important element into your speech that is necessary for the effective communication of your idea, the emotional essence.

EMOTIONAL ESSENCE ("THE SO WHAT")

Along with the literary devices related to analogy, your speech must contain something of yourself. The emotional essence provides personal context for the work you do. I often ask scientists why they do the research they do. Not a few times this question has been met with a blank stare not unlike that of an uncomprehending audience member listening to an overly technical speech about scientific research. "It's my research" is sometimes the answer which isn't an answer to the question, why. "I know it's your research, but why do you do it? What has motivated you to put all this time and energy into the study of Phytoplankton?"

There is a reason why a scientist does the work they do and often the reason is connected to a very personal and passionate interest in something that has real meaning for them. The actor's challenge of answering the question about a character "how is this person like me?" is met by focusing on what is desired by the character. The scientist must do the same and ask themselves why they desire to study something like Phytoplankton.

This is what your speech is about, but once you know what the speech is about, you must look for the meaning of it personally. What is this like for you? How do you understand it in everyday life terms? Why should this matter to anyone else? This is the start of your journey towards discovering the emotional essence of your speech. Ask yourself what it was that first excited you about the work you do. What prompted you to dig deeper into it? How did it make you feel to learn about the subject of your speech for the first time?

What we are talking about here is passion. Though many scientists will frown at the notion of an emotional attachment to a scientific inquiry, a little prodding will soon reveal that they got involved because they had a personal interest in the research and they found it important, exciting, or even fun to learn about their field of study. Some of those scientists will also reveal that they or someone important to them were somehow affected by the research or lack thereof. The scientists I meet in the medical profession often tell of how they were driven to pursue their work because a family member suffered from some condition or disease, and they were determined to discover a cure.

That kind of passion is obvious and can often be understood by an audience with a shared experience, but not all research has the same kind of resonance for an audience. Your passion for your research may be obvious to you, but for our purposes we are challenged to share what we found with others and more specifically to instill in them the same kind of wonder at the subject you have devoted yourself to as your life's work.

When you seek out your own passion for the work you do and begin the process of expressing that passion to others who may not have the same understanding of your work, it is necessary to find the emotional essence. This is the driving force that first hooked you. You are seeking to express it in a way that will hopefully do the same to others. To find that essence you start with your own passion and then ask yourself how it may have the same kind of meaning for others. The metaphors you choose to express your central idea can help with this.

As we have discussed, metaphors are attempts to paint a picture to appeal to the listeners' imagination and to break through jargon to a more shared and common understanding of an idea or concept. But metaphors also invoke an emotional response rather than an intellectual one. This is the essence we speak of now. When you have that metaphor use it to reveal the emotional essence. The emotional essence answers the question: "what is this speech about and why should anyone but me care about it?" Sharing personal experiences connects you to your audience and puts some of the technical aspects of your speech into a more relatable form. You are looking for the "human" aspect of your work to be shared in the moment with a bunch of other humans in the room.

THE ACTOR'S APPROACH

Just as a scientist is charged with finding the emotional essence in their research, an actor is challenged to find the emotional essence of a character that may or may not be like themselves in temperament, history, or circumstance. The particulars of the character's circumstance, who they are, where they are, and what they want are unique to that character. Of the three, the question what do they want is the most important. This is where commonality may be found for the actor. Regardless of who someone is or where they live or have lived in history, to desire something is a universal and very human experience. The actor must ask the question, "how is this person like me?" When you connect to a character's desires you are on your way to finding the answer to that question. We may live in different times under different circumstances, but as human beings we are all driven by the same desires to be connected, to be loved, to be understood.

The actor who is charged with playing a fifteenth century German mystic Abbess may have a tough time understanding that character from a twenty-first century point of view. Not to mention, how do you play a person of German descent who is a member of a religious order and is also a mystic who has visions believed to be sent from God? I will go into this process in more detail in the chapter on Working in Role, but it is worth mentioning here as it pertains to your process of personalizing your speech in order to discover the emotional essence.

Every actor must do research for unfamiliar roles such as the one described above, which by the way is based upon an actual person, Hildegard Von Bingen. She lived in Germany in the fifteenth century and claimed to have visions sent directly to her from God that inspired her to write religious music for the choir. Her music was a means towards her own self-expression. As a woman she was not permitted to speak from the pulpit in the abbey. This was her way of communicating with her congregation.

By using mystical music, which can still be heard today (look it up), Hildegard fulfilled a desire to be heard and understood. Not only is her music still relevant, but her desire to communicate also has meaning for us today in our own quest to connect. Any scientist who has been met with the blank stares of an audience when talking about their research knows what it is to desire to be heard and understood.

STORYTELLING: TRANSFORMING THE CENTRAL IDEA INTO A PREMISE

A further way to develop an effective speech is to make use of the form, storytelling. When you know what your central idea is and you can state it simply and directly, you can then think of it in terms of storytelling. Every good story has a premise, a cause-and-effect statement that is related to the spine or central idea. This becomes the premise for the story you tell to illustrate your central idea.

"Wearing sunscreen can prevent skin cancer." This was the stated central idea for a speech shared by one of my students in a recent course. In this speech, the student spoke of the many reasons why people do not wear sunscreen. Inconvenience was one reason. Some people don't want to take the time to smear that stuff onto their skin. They don't like the feel of it on their bodies or it burns their eyes. It requires a conscious effort to take the precaution of applying sunscreen.

The central idea of wearing sunscreen can prevent skin cancer becomes the premise; "A little discomfort now will save you from a lot of discomfort later."

When you think about the content of your speech it might be helpful to also think about the story you are trying to tell. The facts can be interesting, but to illuminate these facts through the light of a greater more universal truth is the function of narrative.

Storytelling is an ancient and effective form. When you see your speech as an opportunity to tell a story about the work you do, you begin to make a case for why it should matter to the rest of the world. As human beings we crave the experience of community. It makes us feel safe and relevant to belong to something larger than ourselves.

Narrative connects us in ways that mere reporting of facts does not. When you ask yourself about the who, the where and the what of your speech and cast yourself in the central role of a story about your work, you are well on your way to being a better communicator. But to also see others in relationship to you and your central idea, and how together you tell a story that answers a critical question, sends you on your way to the development of a powerful premise.

Ask yourself, who am I in this story about the work I do? Where does this work happen and how does it affect the lives of others? "Wearing sunscreen can prevent skin cancer" becomes "I have met people who didn't wear sunscreen and here is what happened to them," or even stronger, "my brother died because of skin cancer that could have been prevented." If the story is that personal, you may want to consider taking the risk to share it. It may be painful, but it also has the potential to be cathartic for both the audience and you. Either way, ask yourself again, why do I care about this and why should anybody else care as well?

In the same way an actor considers the who, where, and what in their preparation for a role, you must consider these elements as a part of the premise for your storytelling in your speech. To develop the central idea into a premise for your story, there are a few things to consider.

1. *Who* are the characters? This in many cases is you and others related to the premise of your story. *Who* also is determined by role. Characters and people will be a part of a story because they have some job to do or role to play in how the story unfolds.
2. *Where* does this story take place? How does the setting influence the events and people in your story? Often the reason why something happens to someone is because of where they are. Use this element to set the scene for what happens in your story.
3. *What* is the conflict or struggle at the center of the story you are telling? People and places are a part of a good story, but the challenge, dilemma, struggle, or conflict those people are engaged in is what drives the story forward. When you look at the opposing forces in any struggle and ask yourself what is at stake, you begin to build the tension that is at the heart of a good story.

4. *Conclusions* are drawn from each element that add to the overall story-telling. The use of sunscreen or the health of Phytoplankton may not sound like very exciting storytelling until you consider how these things affect people. The harm or good that can come from the choices we make in relation to scientific research is given a sharper focus when the stakes are made clear from a storytelling point of view. If you don't use sunscreen, you may get cancer. The larger story there is one about public health. If you don't pay attention to changes in the numbers of Phytoplankton populations, you may miss the implications this has for global changes in climate. The larger implication of the survival of this ecosystem is related to our own survival as a species on this planet. Each of these subjects contains a who, where, and what as a part of a premise for a good story. We are drawn into the research because we have a stake in what it is about.

Sharing personal experiences is key to your connecting to others. The essence of your speech is concerned with why it should matter to anyone else but you. This is the challenge of sharing your central idea. Storytelling and developing a premise that is connected to that idea is a significant way to make that connection.

Dramatic Structure

As a playwright I have studied dramatic structure and how it functions when constructing a play or screenplay. I say constructing because playwriting is a craft as well as an artform. A play is molded and formed, not just written. Unique to the form is the idea that a play is not complete until it is staged in the three-dimensional theater space. A play must be performed for it to really come to life.

The literary merits of a piece of dramatic writing are secondary to its functionality as a piece of theater. Can you play it? This is the test of an effective piece of dramatic writing. The belief that acting is doing is related to this idea. The actor performs or does what a playwright has guided them to do by providing the elements of dialogue and stage directions. The actor creates behavior guided by the actions described by the playwright. Action is doing.

When preparing your speech, it is useful to think of it as a piece of dramatic writing which will eventually be performed for an audience. You are sharing your speech, not merely presenting it. This is a conversation with your audience that takes advantage of the implied relationship between you and them to create an active event that involves everyone in the experience. Thinking of your speech in this more active way is useful. Preparing it with this idea in mind is essential to your effectiveness as a communicator.

THESIS/HARMONY, ANTITHESIS/DISHARMONY, SYNTHESIS/ NEWHARMONY

When considering the function of dramatic structure, it may help to think of how a piece of dramatic writing moves through the stages of the form. A more specific breakdown is next, but for now we may be able to speak of it this way: A dramatic story begins with a thesis, some commonly understood or accepted state of affairs. This world view is changed in some way by the introduction of a challenge to the status quo or the antithesis to the accepted world view. The conflict or struggle that ensues leads to an adjustment to the status quo that incorporates all that went before, but also allows for a new view of this world, a synthesis of the two into something related but different.

In other words, there is a kind of harmony in the original world view that is challenged by something or someone which leads to disharmony. This struggle is eventually resolved and leads to a new but related world view or a new harmony. The implication here is that the movement from one world view to another is not static but active. Where we start is somehow changed and the process of that change is inherently dramatic.

If once again we use Shakespeare's *Hamlet* as an example, we may say that life at Elsinore Castle in Denmark is in order before certain events change that. Hamlet is the prince. His father, Hamlet senior, is the King and Gertrude is the Queen. Our thesis or harmony is, all is well in Denmark. When Prince Hamlet's uncle Claudius poisons King Hamlet, he challenges the thesis and harmony of the world at Elsinore Castle. His actions lead to an antithesis or disharmony in this world: Something is rotten in Denmark.

It is important to point out that this is where Shakespeare begins his play, at a moment of disharmony. In dramatic writing, this is the point of attack by the playwright. The harmony of the world of the play has been disrupted and so the antithesis or disharmony is what starts the action. Harmony is routine and not very active. To begin at a point of disharmony is active and could be described as simply the breaking of a routine. When a play such as *Hamlet* is started in this way, the original world view or harmony is referenced by the characters' reactions to what has changed. Comparisons are made between the good old days and the present state of affairs. To look back is informative, but the action is already moving forward into the thick of a new reality.

As the play moves forward and Prince Hamlet finally takes his revenge, a synthesis of the past and the present worlds takes place. A new harmony is achieved, but at great cost. Hamlet speaks of the new harmony in his dying words to Horatio,

> O, I die, Horatio!
> The potent poison quite o'ercrows my spirit.

> I cannot live to hear the news from England,
> But I do prophesy th' election lights
> On Fortinbras. He has my dying voice.
> So tell him, with th' occurents more or less
> Which have solicited—the rest is silence.

In other words, my work is done here. My father's death has been avenged. I can die now, and a new generation will take up the fight. Hopefully they won't make the same mistakes.

The understanding of this part of dramatic structure can be useful when writing your speech. Your research begins with a thesis that points to a certain reality or harmony in the scientific world. Harmony doesn't necessarily imply that all is well. You may be just stating the way things are when you start out and how your research is concerned with challenging some of the more traditional views on the subject. In that sense, the antithesis or disharmony you present is meant to shake things up and hopefully lead the world to a new understanding, a synthesis or new harmony of a sort. If you can do that, "the rest is silence."

Elements of Dramatic Structure

There are five basic elements to the form of dramatic structure including exposition, rising action, climax, falling action, and denouement.

Gustav Freytag, the nineteenth century German playwright and novelist, expanded on these elements and created a pyramid to highlight seven parts he considered necessary to storytelling: exposition, that contains an *inciting incident*, rising action that contains a *complication*, climax, falling action initiated by a *reversal*, leading to a *resolution*, and denouement. Here is an image and explanation as described by Janice Campbell on the Excellence in Literature website (Campbell 2023).

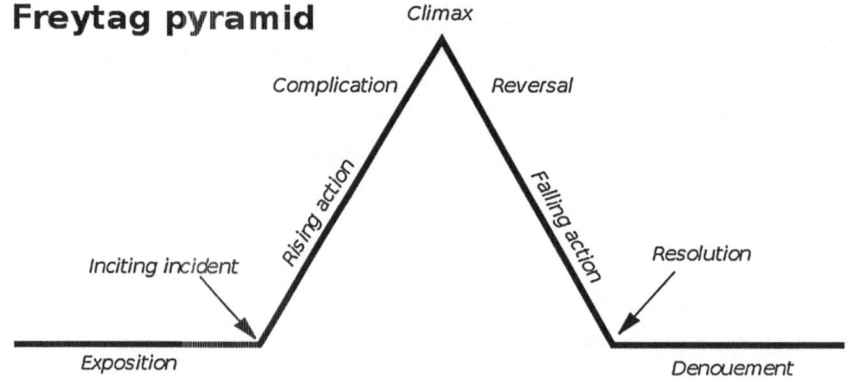

Exposition: The storyteller sets the scene and the character's background. (Harmony)

Inciting Incident: The character reacts to something that has happened, and it starts a chain reaction of events. (Disharmony)

Rising Action: The story builds. There is often a *complication*, which means the problem the character tried to solve gets more complex.

Climax: The story reaches the point of greatest tension between the protagonist and antagonist (or if there is only one main character, the darkness or lightness of that character appears to take control).

Falling Action: The story shifts to action that happens because of the climax, which can also contain a *reversal* (when the character shows how they are changed by events of the climax).

Resolution: The character resolves the problem or conflict.

Denouement: Happening simultaneously with the resolution this is the unknotting or unravelling of things. Specifically, all the conflicts and complications that made for our rising action now come undone as certain truths are revealed and we rush towards the ending of our story. The denouement is often happy if it's a comedy, and dark and sad if it's a tragedy. (New Harmony)

You can access this form and apply it to our work in writing speeches about your science research. If we go back to our original notes on dramatic structure you can see how these elements work together to form a whole.

The storyteller, now you, sets the scene in the exposition. You begin by telling us about your work as it relates to the established field of study for your science. In this there is a thesis or harmony that describes the way things are now. Remember that harmony doesn't mean that all is good. There may be a problem from the start and your work is concerned with addressing that problem. Your work may be seen as an inciting incident.

In the inciting incident, you now tell us of your reactions to the current state of affairs in your work as a scientist. Your reaction to the thesis or harmony of currently held practices or prevailing views in certain scientific research may be presented as a challenge to the status quo. A new way of thinking about a scientific subject can be daring in and of itself. To offer up a challenge to commonly accepted theory or practice you will have to be prepared with evidence to support your research. It can be dramatic, but it must also be true and supported by facts. The disharmony that results from this action on your part needs to be supported by an antithesis that has been carefully researched by you.

If you do have a legitimate case to make that challenges some established views or practices in the field, then you will become a catalyst for change. In the context of dramatic structure, this part of your speech adds a complication to the rising action. Here is where you get into the heart of your central idea. You have prepared the audience for the more challenging aspects of your research, and you now create a more complete picture of just what is happening in your field and what else needs to be done to move your research forward.

If you commit to this moment and play for keeps as the actor does when playing a role, you will be inviting the audience to take a serious interest in your work by involving them as a participant in the conversation. You have also raised the stakes with a specific challenge that is supported by evidence. Your seriousness of purpose is demonstrated by your conviction to an antithesis that has been well researched and well-rehearsed.

What follows is the climax of your speech. Here you bring together all the elements of dramatic structure as you point the audience towards a new understanding of the scientific subject, the central idea of your speech. This synthesis of the old and new is meant to be shared with your audience in a collective understanding. Hopefully your speech has instilled a sense of connection or ownership of the problem in the audience so that they see themselves as being actively involved in the challenges offered by your speech and not just passive bystanders.

Once the climax has been established, your speech moves forward towards the denouement. Here the action begins to fall, but a resolution to the issues you have discussed is also a part of the experience of the speech for you and the audience. It is important to note that in dramatic writing, resolution is not the same as solution. To solve a problem is good, but when you add the re: to the beginning of that word, meaning *again*, you are suggesting that the issue is ongoing and is something we should continue to explore. This idea, like science itself, is active and ever changing.

When you commit to being open and curious about the future of your work, you are inviting the audience to do the same. The denouement of your speech contains in it a prompt for the audience to consider a new harmony for the work going forward. In that new harmony is the suggestion that the process goes on as we face the challenges of your central idea and of science together.

STORYTELLING AND PLOT

One last distinction to make here as you consider the writing of your speech. As listeners, when we are told a story, often we want it to have some chronological order and an unfolding of events that hint at an overarching meaning to the story. The same is true of the story you tell in your speech. "This is a speech about *what* and it has meaning *because*." To find that meaning it is useful to consider plot as a part of storytelling. The term plot is often used interchangeably with story. They are related, but not the same. A story contains a plot which helps to activate and connect episodes or events. In that sense, plot is an essential part of a dramatic and active story.

A story may be described as a series of incidents containing characters (who), settings (where), and conflict (what). These elements alone can make up the functioning parts of a story, but they don't really move the action forward without plot. A story can be expressed as this happened to these characters in this place, then this happened, then this happened, etc. Plot takes those same elements of a story and connects them. This happened to these characters in this place and then *because* of that, this happened. The cause-and-effect sequence is a way to describe what plot does. Events are connected and so they depend upon each other to move the story forward. This is also related to the pinch/ouch sequence in the actor's process. Remember that?

If you look at Freytag's pyramid above, you can see that the term inciting incident is just another way of saying here is where the plot begins. The complications or questions that cause the action to rise are where the plot thickens. The climax is where the elements of plot all come together. Here all the plot points or questions in the cause-and-effect sequence have been posed and resolutions or answers to those questions have been revealed. Once that happens the ending or denouement is all that is left. There is nothing left to advance in this story.

PREPARATION IS ALL

I can't stress enough how important it is to work on the actual writing of your speech before you ever think about sharing it with an audience. Stanislavski's book, *An Actor Prepares*, is concerned with how an actor gets ready to play a role. There is much work to be done before that performance ever happens. So too does a playwright work to understand the story they are telling with research, rewrites, and revisions. Only after they have made and remade their script many times over do they consider showing it to a producer in the theater.

As a communicator of your work in science, you are charged with the tasks of both writer and performer. You work to prepare the best, most complete speech you can write and then you work again to prepare for the sharing of it with an audience. The best way to have a positive performance of your speech is to make sure the speech itself is the best it can be. The best way to ensure that is to work hard at preparing it in the first place.

CHAPTER 7

Analyzing Your Speech

Abstract The next step in the development of an effective speech is analysis. This chapter looks at the first steps in preparing your speech for sharing with an audience. Breaking down a speech into units of action or beats and choosing objectives to help you to set goals that are doable and achievable is also covered in this chapter.

Keyword Speech analysis

TEXT ANALYSIS: FIRST STEPS TOWARDS SHARING YOUR SPEECH

Just about every traditional approach to rehearsal process in theater begins with an analysis of the script. We will speak more specifically about the rehearsal process in another chapter. The focus here is analysis as a part of preparation of the speech. Like the script of a play, your speech can benefit from the process of text analysis. In the theater world this process is an essential part of the work of a director in preparation for the rehearsal process.

Before a director meets with a cast of actors, there is much work to be done on a potential play script. This is a solitary process done months in advance of any formal rehearsal of the play. So too are you charged with the preparation of your speech before you find yourself sharing it with an audience. As the writer, director, and ultimately performer of your speech, you do well to devote some time to the preparation of your speech with an eye towards performance.

The theater director first spends time alone working out the structure of a play. What is this a play about, who is the central character, what is wanted by that person and from whom? These are the first questions a director asks in that initial phase of script analysis. Further process leads to a consideration of a central metaphor and perhaps an understanding of the style of the piece.

D. Dannenfelser, *The Art of Effective Science Communication*,
Palgrave Practical Guides in Communication,
https://doi.org/10.1007/978-3-031-57030-8_7

You can see the similarities here. First you spend time writing your speech and asking yourself some of the same questions. What is my central idea? This is a speech about what exactly? Who are the characters? These are the people, yourself and others who share a concern or are connected to the struggle expressed in the central idea. You may also be the central character, the one most concerned with the struggle. What is that struggle, the thing that prompts you or others to take action? All these questions help to lead the director to the development of an approach to the play. These same kinds of questions lead you as the speaker to an approach to your speech. The approach is the practical consideration for the *how* of the play or speech.

To further the analogy of this process in theater let's look at the roles of others. The point of view of the playwright is discovered through careful analysis of the play. This point of view leads to a vision of the play in production by the director and eventually leads to a choice about the approach that director will take to see that vision come to life on stage. In the actor/director relationship, the actor is responsible for delivering the character in the play while the director is responsible for delivering the play itself in a production that considers the intentions of the playwright, the actions of the actor, and the imagination of the designer.

Of course, there is crossover in all these roles. The actor also makes use of the imagination. The playwright also makes choices about the actions that will become objectives for an actor and the director looks at the whole and decides upon a point of view and approach that encompasses the work of everyone involved. As the writer, director, and actor for the sharing of your speech, you do all these things. It benefits you to take some time to seriously analyze the speech you have created before moving on to those other roles related to performance.

Breaking Down Your Speech into Beats

Once you have a final draft of your speech and you are happy with it, you now begin the process of readying that speech for sharing with an audience. The first step is to read over the speech just to listen to the story being told. Take your time and just read what is there without judgment. You have already done this many times while writing the speech, but now you are reading with performance in mind. It's not just about what you wanted to say. It's also about *how* you want to be heard. Again, the *how* is very important to your overall approach. The *how* is active and full of doing. And as I have said in the chapter on Acting Techniques, you must always be doing.

After that first read through focusing on story you read the speech again. When you read the speech this time, you are listening for the changes in direction, the transitions between your thoughts about your subject. You are looking for the movement of the speech within that larger structure of the dramatic action of storytelling. Every time you make a transition from one point to another something changes. That change is a beat.

What is a beat exactly? Simply put, a beat is a unit of action. For example, when examining the structure of a novel or a short story the novel is broken into chapters which in turn are broken up into passages or scenes. Each of these can be looked at as a beat or unit of action. A play too has divisions or units of action. A full-length play is divided into acts which are further divided into scenes. Within these scenes are specific units of action which are interpreted by the actor and referred to as beats. For the actor these transitions within a single scene lead to choices that may be made about the behavior of the character tied to specific objectives.

One major difference between certain forms of storytelling like novels and other forms such as plays is the tense. Novels tell stories that have already happened in the past. Plays tell stories that are happening now in the present and are being witnessed as they unfold in real time before a live audience. It is that immediacy that we are seeking when sharing speeches even though we may use stories in the past tense to illustrate a point within the speech. We also are attempting to share a discovery in the present with the audience in real time as the speech is happening, right now. A tricky bit of work, but possible when you understand the work of the actor. Here is some guidance more to that point and related to our discussion of dramatic structure.

When you begin your speech, you start by introducing the central idea in the form of some challenge, problem, or conflict. You also identify the characters who are involved in this conflict in this first beat. Often it is you who are cast in the role of central character as it is you who have done the research that has led to this moment when you will share what you have found. In the form of dramatic structure this is also known as exposition. We can refer to it as the first beat.

After having sufficiently set up your science story, you then get into more detail and support that original idea. Keeping dramatic structure in mind, we can think of this as the rising action of your speech where you may add a complication to the original conflict deepening the challenge and opening the possibilities of how to meet that challenge. Remember the key here is to build the tension in your story. That becomes a second beat.

Moving forward you have now built to a moment where the tension is the highest in the conflict contained in your story. This is the climax or climactic moment where the action of the conflict in your speech/story reaches its peak. Again, this is another beat.

When you move forward from here you are in the territory of resolution. The falling action of this part of your speech is focused on resolving the conflict in your story and bringing your speech to an end. This is the fourth beat which is quickly followed by a conclusion or denouement to your speech. This may also be a fifth or final beat.

Five beats are about average for a short speech, about three minutes in length. If you are sharing a longer speech, you may find other expansions on the central idea that lead you to include more beats, but five beats are about right for you to cover the basic dramatic structure. At least it is a good way to get started. You can go back and adjust and refine the transitions to discover more beats within the beats, but this basic way of organizing your speech will help you to make those decisions in further revisions.

The Central Idea or Premise Related to Choosing Objectives

Once you have done your analysis of your speech and have discovered those units of action we call beats, you now must go back and decide just what the action is about in each one of those beats. In other words, you will start the process of choosing objectives for each of your beats. An objective is an action, and these two terms are often used interchangeably in the actor's process.

As we have discovered in the process of writing your speech, you are searching for a central organizing idea that holds your speech together and drives it forward to some overall point you want to make or a point of view you want to express. This central idea also can be viewed as the premise for the story you are telling in your speech. Like the spine in your body, the central idea is at the core of everything and the objectives you choose are directly related to that spine.

As you look at each beat, ask yourself what is literally happening there. For example, when you first begin your speech, you may start by introducing yourself. If you have been introduced, you may launch into an introduction to the problem or the challenge your speech is meant to address. This is your central idea and everything you say or do is somehow related to it for the duration of the speech. At any rate, you begin your process by asking yourself what it is you are literally doing in that first beat. I am introducing myself or I am introducing the central idea.

You then decide what you want to do in each beat related to your Central Idea. Remember that every choice you make is related to the central idea even if you are simply introducing yourself, you are doing that with the central idea in mind. That is the reason you are here to speak today after all, and you are the person who can best speak about the central idea so here you are. But what do you want? This is the next step. To answer this question, you must choose a *verb* that captures the essence or meaning of the literal action in the beat. Let's speak more specifically about that process for forming objectives.

Forming Objectives: A System of Wants

The partners in an actor's work on stage are the fellow actors' characters in relationship to your character. For our purposes, the term partner refers to the audience for your speech. Either way, the formation of a strong, active, and specific objective for each beat is essential if you wish to create a dynamic sharing of that speech.

So, what is an objective exactly? The great director William Ball tells us about the form for objectives in his book, *A Sense of Direction*. He describes the objective form as "A system of wants" (Ball 1984). He further describes this system using a form. The form looks like this:

VERB	RECEIVER	DESIRED RESPONSE

Applying this form to our task of public speaking we can ask two simple questions, what am I literally doing and what do I want? In each beat you must know what the beat is about and then decide what you want to accomplish with it.

For example: In Beat #1 you are literally introducing yourself to your audience, but objectively you are *assuring* your audience that you really want to connect with them today. In other words, I am literally introducing myself, but objectively I *want* to *assure* (verb) my audience (receiver) that I'm here to connect (desired response). In Beat #2 you are literally presenting the problem, but objectively I *want* to *awaken* (verb) my audience (receiver) to their sense of outrage or concern about my central idea (desired response).

The *want* part of this phrasing is important. It activates you and gets to what you desire to *do* in the beat. Acting is doing, but it starts with wanting. That said, you can only really act upon the first two parts of the objective form. The verb and the receiver have all the action contained within them. You are choosing to *do* something (the verb) in relation to your partner (the receiver). Here is where you can act. The last part of the objective form (the desired response) is not actable. This is a hoped-for response. It has the goal of the objective contained in it, but ironically you have no control over this part of the form.

This harkens back to the idea we touched upon earlier concerning result orientation. Your focus is on doing and being present in the moment. This translates to pursuing a clear and specific verb in relationship to a receiver or your partner (the audience). Whatever happens next is out of your hands. But if you have done your preparation and have developed a strong and deliberate objective for each beat and you are busy and engaged in pursuing that objective in the moment and with a hundred percent commitment, the desired response should take care of itself. This is where that element of trust is most pronounced. If I do my part, the desired outcome will follow.

What If I Don't Get the Response I Wanted? Making Adjustments Not Judgments

But what if the desired response doesn't happen? What if I do my part and really pursue a well thought out objective with an active verb that keeps my partner at the center of my actions? What if I really do that and something other than the desired response happens? Your first thought may be that you are doing something wrong. But if you remember our earlier point about manipulation, this is where you have to allow for the unexpected response and accept that in the moment. Consider this; the actual response you get may be more in line with what your objective was about than you had anticipated and so may teach you more about what you wanted to share than you had known yourself. In that instance, you may have discovered a different truth about your subject, your audience or yourself and that could be a revelation. Should that happen, you say, "thank you" and incorporate that discovery into the objective form going forward.

If that response is not so positive or even creates a negative outcome, you are still not wrong. You simply need to go back and make an *adjustment* to your objective. The objective is at the root of every action, the driving force. If it takes you down a road that is confusing or way off the mark in relation to your central idea, then it is time to go back to the root and reform the original objective to be better in line with what you had intended. Either way, you are trying to refrain from judgment by getting busy addressing the challenge through *adjustment*. Your own response to such a challenge therefore should always be adjustment not judgment.

The Central Idea as Super-objective

We have referred to the central idea as the organizing thought, the thing that you want to talk about above all else in your speech. We have also spoken of the central idea as the premise of the story you want to tell in your speech. We may also think of the central idea in terms of objective. But the central idea is not just an individual objective related to a single beat. The central idea is the overall objective or super-objective. This is the objective that all individualized objectives stem from, the spine of your speech.

Stanislavski puts it this way, "In a play the whole stream of individual, minor objectives, all the imaginative thoughts, feelings, and actions of an actor, should converge to carry out the super-objective of the plot. The common bond must be so strong that even the most insignificant detail, if it is not related to the super-objective, will stand out as superfluous or wrong" (Stanislavski 1936). In other words, if any part of your speech is not somehow related to the central idea or super-objective, you must cut it. This agreement you make with yourself will lead you to a dynamic and specific sharing of the content of your speech. Just like how every bone in your body is held together centrally by your spine, every objective you choose to activate your speech for performance must be connected to the central idea or super-objective.

CHAPTER 8

Working In-Role

Abstract The power of working in-role is a concept explored in this chapter. How playing the role of speaker can be compared to the actor's work of playing a character is discussed in detail. The freeing nature of this process is discussed and basic structures for how to use this technique are outlined here. Examples of role-play and role-taking are offered from the arts as a demonstration of the concepts of the distance model and projective technique. All are connected to the task of public speaking for the non-actor.

Keyword Working in-role

There are, of course, many books on communication and, as I've mentioned, public speaking. These, however, are not necessarily the same thing or more specifically, they should be connected, but aren't always. There are many people who find themselves in the position of having to be a public speaker, but do not in any way think of themselves as that. Often, it is the last thing they would choose to do if given the choice. For these people the thought of standing up in front of an audience and somehow capturing their attention seems an impossible task.

When we consider the concept of working in-role, public speaking can be seen as a task, a job, a specific event that is structured and has rules for achieving a specific outcome. The task of public speaking, though daunting to some, can be accomplished just like any other task if this specific structure is understood and more importantly prepared for by the speaker. *An Actor Prepares*, that essential text written by Stanislavski makes this process clear. It is a system, as he calls it, whereby the actor and for our work the speaker can approach a task with an objective and pursue a desired response or specific outcome. "Whatever happens on stage must be for a purpose, not merely the general purpose of being in sight of the audience. One must earn the right to be on stage and it is not easy" (Stanislavski 1936).

But what if the mere thought of standing up and addressing large or even small groups of people makes your stomach turn? The actor refers to this as stage-fright. Others just call it nerves. Either way, the assumption that anyone should be able to overcome this feeling doesn't allow for the very real differences that occur in people. The introvert is different from the extrovert. The less gregarious among us would much rather focus on research done in a lab, in the case of a scientist or in my case, the text of a play written in my basement office away from the public eye than subjective themselves to public events that make them feel nervous or exposed in some way.

People are different. The choice for some to seek out a career that involves quiet study and careful research rather than the very public position of spokesperson is an indication of this reality. Some people are show horses and some people are work horses. Even so, in our ever-escalating public access society, the requirements for communication are affecting more and more people and the work horses are sometimes called upon to explain their work or even defend it in the face of stiff competition for scant resources. If you make a discovery of any significance, chances are you will be called upon to share that information with the public perhaps to guarantee funding and to justify going forward with your work. But being at the center of this kind of attention just isn't you. So, what do you do?

To work in-role is to understand the use of character. This is obvious enough for the actor. There are many introverts in the performing arts although you would never know it when you witness the work of these people. How do they do it? Character and the understanding of it may be the answer. When an actor approaches a role, they ask questions; how old is this person, where did they grow up, what are their social or political views, their economic reality, or their psychological state of being? The background or beyonds of a character determine their world view, but more importantly they determine their behavior. An actor makes choices about how a character will behave based on where they come from and their world view.

Here then is a technique for accomplishing a task that requires you to stretch beyond the limits of your usual comfort level. When you put a character to work for you, you don't lose yourself. Rather you discover another aspect of yourself that is more comfortable with the demands of a public appearance. Working in-role means you employ the specific structures of projective technique to accomplish your goals. When you embrace the introvert, you essentially use what you understand about yourself to play another.

Kurt Vonnegut and Working In-Role

Kurt Vonnegut was a very famous author of many novels and short stories. I will talk more about his gifts as a public speaker in the chapter to come on rehearsal. For now, I want to tell you about a short story he wrote that later became a short film entitled *Who Am I This Time?*

When preparing this chapter, I thought back on this story because at its core is an essential understanding of the concept of working in-role. Of course, I had no idea what it meant to work in-role when I first read this story. It wasn't until later after taking up my training as an actor and director and then studying these concepts at NYU with my professor, Robert Landy, that I came to understand the significance of Vonnegut's ideas on this. Working in-role is an important part of the approach I have developed for anyone interested in public speaking. This is the theory behind the Distance Model and the use of Projective Technique. More on that to come.

In Vonnegut's story, the character Harry Nash is a very shy man who works as a clerk at a hardware store in a small mid-western town. To most people, most of the time, Harry is invisible. He keeps to himself and does his job making little to no contact with other human beings, preferring to stock the shelves at the store rather than engage with others. Most of the time, this is Harry's routine; go to work, keep to himself, and just get along unnoticed and alone.

I say most of the time because there is one exception to this routine. For some reason Harry is drawn to the local theater group and regularly auditions for the plays they perform. And here is where the story really begins because Harry doesn't just audition for the roles in these plays, he nails it every time and usually lands the lead being a kind of savant when it comes to acting.

Harry, in an audition and subsequently in a role in a play, is not the shy, introverted hardware clerk of his true-life reality, he is King Lear, Willy Loman, Stanley Kowalski, or any other major character he needs to be when in rehearsal or in performance. By the way, if you don't know who those characters are or what plays they appear in, shame on you. Stop reading now and go and look them up. I'll be here when you return.

The amazing idea expressed by Vonnegut's creation of the character Harry Nash is that a person can find certain capabilities within themselves through an exploration of the roles they may play. For the main character in Vonnegut's story, the structured, fictional reality is a form that allows for Harry to be transformed. Like every good writer, Vonnegut understands that this premise is not enough in and of itself. He knows that Harry's other routine must somehow be challenged to create conflict in the story and interest in the reader.

Enter Helen Shaw, another character who is visiting the town for a while and who decides to try her hand at the local theater group herself. When she meets Harry during an audition and mistakenly believes he is who he is playing, the plot thickens, and a remarkable turn of events follows. The ending is a great example of the use of Role-Playing and instructive to anyone learning about this powerful technique. I won't spoil it for you here. You must read it or if you prefer, see the film with Christopher Walken as Harry and Susan Sarandon as Helen.

For our purposes I will say this, working in-role is not about becoming someone other than who you are. It's about discovering other aspects of yourself that you may not have access to in your day-to-day existence. It's about finding the public speaker in you and using the form to allow that person to emerge by simply identifying the structure of the task before you and seeing your role in it.

DIE AND TIE

In his book, *Drama Therapy (Concepts and Practices)*, Robert Landy explores the uses of drama in the treatment of people suffering from trauma, insecurity, and other psychosocial issues. Dr. Landy was my professor when I was studying for my Master's in Educational Theatre at NYU. His course on Drama Therapy was a significant part of my work in that program and I have used many of the concepts I discovered there in my work as a teacher and an artist. While we do not attempt to conduct therapy as a goal of this book, ours is a mission concerned more with communication, we recognize the benefits of the use of drama and understand how these techniques can be applied to that objective.

To understand how the concepts and practices of drama therapy may be of use to us as communicators it is important to first understand the difference between DIE or Drama in Education and TIE or Theatre in Education. Simply put, DIE is concerned with the process of drama and its implications for the better understanding of self or others. DIE has provided a useful set of processes for individuals working through personal issues, but it has also provided a framework for educators whose work with students has been enhanced by the forms of role taking and role play that are essential tools of DIE.

The great British educator, Dorothy Heathcote, made ample use of these and other practices in DIE to teach students in many subject areas. A history class for her students involved direct engagement in a kind of living history. Her students didn't just read about the Mayflower and the pilgrims' voyage across the ocean, they built their own Mayflower out of desks, chairs, and found objects in the classroom and became pilgrims themselves, each developing a specific character with a beyond or story that gave them a reason to be on that ship in search of a new world.

Cecily O'Neil, another leader in the field, explored an aspect of US history in the story of the Oregon Trail. Here too her students developed detailed personal histories for their characters each one having a very specific backstory, but more importantly a specific reason for leaving their homes in the east to travel west and into the unknown. The form is well known now and can be witnessed in the programing of The History Channel and others.

The concept of teaching here is immersive. You don't simply read about and discuss the events of the past; you investigate the lives of people from past cultures and then you play those people through reenactment and role taking. For the student this is an active and direct experience of the "other." Their under-standing of someone else's existence is enhanced by an attempt to walk in that person's shoes. The basic structure of who, what, where found in every theater event is used in the classroom to offer up a living history. The students become characters in a story of their own making guided by research into the lives of others.

The subject of History is an obvious choice for the employing of these tech-niques, but Science programs on television have also moved in this direction often focusing on the who, what, and where to tell a story about the work scientists are doing. Simply relaying the data is not enough for our media-driven society. We crave the context of narrative to make sense of the informa-tion. Our need to personalize our experiences has made it even more essential that we see ourselves somehow in the science or the history being explored.

"Why does this matter to me?" is the question everyone asks these days when presented with a series of dry facts and figures. The answer is found in that universal experience of identification that comes with the enactment of another's reality and then making the intuitive connection to one's own exis-tence. Identification with another's plight or another's point of view leads to empathy for that other. This is an exploration of self through the identification with another, an essential aspect of the distance model. More on that in a moment.

TIE is related to DIE, but the goals are somewhat different. Where DIE seeks to explore the meaning of self by relating to the experiences of others, TIE is more of a celebration of the different forms of art that have enriched the societies of the past and continue to do so today. Theater in Education pro-grams are structured to provide experiences in the arts for the sake of the arts themselves.

Our cultural growth and enrichment go hand in hand with our development as people and like DIE, TIE uncovers those aspects of the human condition that we all share. Also, like DIE, we are given an opportunity to gain insight into ourselves through the art of others. TIE presents the artistic event as a demonstration and realization of the human spirit in action. For educators these performances also prompt a response in the student to identify with the art form, to see themselves in it and to learn more about themselves by witness-ing the stories told by and about others.

Another professor of mine, Paul Austin, used to speak about the purpose of theatre and for that matter, drama in society as being two-fold. In his view the purpose of theatre was to illuminate and to celebrate. DIE and TIE can be looked at as the forms used to do just that.

THE ACTOR AND THE ROLE

How then do we use these dynamic forms for our own purposes? This is where a more organized and deliberate system is needed to harness the power of drama and apply that to our goals as communicators. As I have mentioned earlier, character is one way that an actor can use to find a way into enactment of a role. Specific biographies for fictional people make them more real and allow for the actor to begin the process of first understanding and then playing that person in a drama.

When an actor first approaches a character, they start by asking questions that investigate how the character is both alike and different from themselves. This initial work on identifying aspects of a character leads to an understanding of that person's desires. The actor then compares those desires to their own to find where they may need to do more or less work to understand what the character may do to achieve their goals. The actor is looking for a way into the character.

Once they know a bit more about what is wanted by the character and they have calibrated the similarities and differences in their own temperament and desires, they can then start the process of making choices about objectives that lead to tactics or behaviors the character may engage in to get what they want. This is a deliberate, but also delicate process. When the actor makes comparisons between the life of a fictional character and their own life, they are engaged in a process of self-examination. This process requires the actor to make a distinction between self and role.

In this initial phase the purpose of the study of self is to engage in an examination of your own uniqueness. Your personal understanding of how you are different from others is informed by your experience of others in relationships. For the actor, a character can prompt certain untapped aspects of their personalities to come to surface or as the psychologist Carl Jung puts it in his book, *Memories, Dreams, Reflections*, "In each of us there is another whom we do not know" (Jung et al. 1961). Self in this early exploration is viewed differently from role. Self is who you are while role is what you do. This is useful to the actor, but also to the non-actor who is working to become a better communicator.

As children we experiment with playing different roles. We witness the behavior of significant others, such as a parent or guardian, and copy or mimic those behaviors. An important aspect of this process that has significance for the actor, the non-actor, and for human beings in general is that by mimicking the behavior of others we grow in the understanding of ourselves. Copying others helps us to develop our own unique selves. We do not become the other, we become more of ourselves by developing our understanding of the roles each of us play in our lives.

When an actor takes on the role of a character, they create a physical manifestation of a person that involves a particular way of moving or speaking. These are influenced by physical limitations or attributes of both the actor and the character. A synthesis of the artist and the fiction takes place which is a demonstration of the actor's skills for mimicry and grounded in emotional truth. But even the most seamless portrayal by an actor of a particular character will not cause that actor to become that character. When we witness a performance by an actor and are lost in the reality of the fiction, we do not actually believe that the actor has become that character. It is our willful suspension of our disbelief in our desire to have an emotional experience of our own that allows for the illusion of reality to be complete and satisfying. Just as the actor goes through a process of identification to play a character, we likewise identify with the actor and appreciate their skill at convincing us to believe in the fiction for the moment.

Ultimately, we are both engaged in a process whereby we collectively seek the truth. But as I have said in an earlier chapter, dramatic truth is not necessarily the same as actual truth. The reality of the fiction helps us to see into a story and find something there that teaches us about ourselves. We enter into an agreement as artist and audience to go on a journey together to discover something more real than the fiction we use to uncover it. Or as the saying attributed to Mark Twain goes, "never let the truth get in the way of a good story."

So how then does this idea help us to be better communicators, or at least to feel more comfortable in our role as such? To be successful as a communicator it may be useful for us to look at how role-taking functions. When we take on a role such as speaker, the role acts as a kind of mediator for the experience. This is true for both us and the audience in the experience. Role-taking is a useful and productive exercise for our everyday lives. The builder builds, the doctor tends to the sick, the teacher teaches. Each of these, and many others, are roles that have clearly defined parameters for functioning.

But these roles also allow us to accomplish some challenging tasks, tasks that without the structure of a role may be much more difficult and even impossible. Again, the effects of the imposter syndrome take hold. How can I take on the authority of a professional when I feel like an imposter in that position? What have I done to give myself permission to tell others what to do? This idea has come up many times in my work with first year medical students. This is a huge hurdle for many of those students: "Upon acceptance and commencement of my medical career, I am now expected to take on the role of developing doctor while still seeing myself as a student." The feelings of inadequacy, given the demands of such a role, are rampant in the medical student community. The bar is set very high, and many feel they not only can't reach it, but don't have a right to it. The young med student is in a struggle to become that which they can only imagine.

This is just one example, as there are many of you who can relate to these feelings of inadequacy in the pursuit of your own professions. But I bring up the medical school model because of the intensity of the work and the particularly high set of expectations of performance necessary for their success. The competition is fierce. And here is the irony; competition is the last thing a student should have as a concern when working towards a successful career. If you remember our discussion about objectives, the desired response is the part of the form that is concerned with results. You may also remember that by committing to the form, you are agreeing to let go of a result orientation. You have no control over the desired response. Your focus is on doing. But how do you keep doing when you don't believe you are good enough? Consider the distance model.

The Distance Model

What is the Distance Model? As described by Dr. Landy, "Distancing is that of an interaction or intrapsychic phenomenon characterized by a range of closeness and separation" (Landy 1986). What Dr. Landy is talking about here is balance between our emotional selves and our intellectual selves. If we think of this concept using the metaphorical image of a scale the graphic might look like this.

Under distance	Aesthetic distance	Over distance

∧

One end of the scale represents the concept of under distance. Here is a person who is very close to the emotional experiences of life. This is an intuitive feeler. This is a person who acts on instinct and experiences the world through an emotional lens. Even so, when we speak of the under distanced person, we are referring to someone who is at an extreme point of functioning. The emotional life may be full of feelings for this person, but they also experience those feelings at an unhealthy level that is imbalanced and can affect their daily functioning in the world.

The over distanced person, on the other hand, is very far removed from the emotional life. These are the thinkers. They are analytical and rational. But like the under distanced person, this is an extreme. The rigidity of their perceptions causes them to see themselves as separate from others, only connecting through thought, and often that thought is reflective of their own views. Thought itself is more important than feeling, or as Landy puts it they experience "a separation of thought from feeling" (Landy 1986).

Under distanced and over distanced people are present in our society, but more commonly people are engaged in a struggle to reconcile the two states of being. This is an effort to find balance. Most of us exhibit characteristics of "both and." At different times, and under different circumstances, it is good to be one or the other, but mostly it is best to be a combination of the two. When you can function by recognizing these aspects of your nature and are consciously working towards that balance, you will be on your way to finding aesthetic distance, a healthier way to live and to experience the world around you. For our work in the arts and as communicators, understanding this balance can be very useful and help us to succeed in our respective challenges.

Projective Technique: Uniforms and Props

It may be useful to consider the example of the medical student once again, to better understand how projective technique works as a part of the distance model. When faced with the challenge to deliver bad news, such as is required of a medical doctor at times, we draw upon the power of the role. The white coat and stethoscope are not merely uniform and prop. They are significant elements for the playing of the role of "doctor," giving people the authority to do difficult things. The power of the role gives us the strength to go beyond our own human frailty. The uniform signifies that authority, and the stethoscope is a tool reserved only for those who have achieved a certain level of mastery in their field and know how to use it. You wouldn't hand a chainsaw to someone who's never even seen a tree before, or as Tyrone Guthrie the great theatre director once said, "the stage is like a tightrope. You wouldn't step out onto it without some serious training."

The actor puts on a costume and handles props, but only after having gone through a rigorous rehearsal process whereby they have come to fully understand the character who dresses in that way and uses those certain props or tools to go about their daily business. These things are not substitutes for knowledge or experience, they are signifiers of a level of competence in a person's training that say, "I am not only ready to do this job, but I've earned the right to do it." A uniform is an indication of a person who has paid their dues.

But there is more to projective technique than just signifiers of authority like a uniform or a prop. Projective Technique can be used to help you or others through difficult or challenging circumstances. To project onto something is to create distance, as we have examined. To give yourself space through the structure of a role, the wearing of a uniform, or the using of a prop, is useful, but the receiver at the point of your communication efforts, the audience, may not be so easy to convince of your authority. You will have to embody that role with a clear and specific understanding of its intentions. As a public speaker you prepare not only to dress the part, but to play it.

Examples of the Distance Model and Projective Technique at Work in the Arts

The famous playwright Bertolt Brecht was an avid practitioner of the form known as Epic Theatre. Though the form had been established earlier by other theatre theorists such as Erwin Piscator, Brecht further developed it into a more unified approach. A particular aspect of Epic Theatre made clearer by Brecht's work was something called the Alienation Effect. The focus here was on thought rather than feelings. This may sound odd when talking about theatre. The word dramatic has been used to describe overly emotional people who quite often act before they think. So, what business did Brecht have in encouraging the actors and the audience to put the drama aside and think about the ideas behind the actions in the play?

Brecht was interested in social reform. His goal was to instigate change in society, to illuminate injustice, and call the audience to action. This was a personal choice for him. Having been targeted by the Nazis for his views, he was forced to flee Germany to survive. But survival was not enough for him. He wanted to challenge the social structure and belief systems that made it possible for the Nazi party to exist in the first place. Brecht didn't want to simply make the audience feel something about the state of the world around them. He wanted to prompt them to do something about it, to act. To accomplish this, Brecht employed techniques of the Epic Theatre and the Alienation Effect.

His plays, at times, were characterized by a documentary style, opportunities for audience interaction, actors playing multiple characters, and transitions in scenes in full view of the audience. There was no attempt at disguising the mechanics of the stage. Brecht's characters would break through the imaginary fourth wall between them and the audience. They'd address the audience directly with declarative speeches or narration of the events on stage. The use of the stage in this way was commonplace in Epic Theatre. These techniques were not only employed by Brecht, but by many other theatre practitioners throughout the ages. Elements of the Epic Theatre can be found in many forms throughout theater history, from the Greeks to the protest plays of the 60s, and to the present day.

In *Waiting for Lefty*, a play inspired by the 1934 New York City taxi strike, the playwright, Clifford Odets, makes his political views known. Influenced by his membership in The Group Theatre, a politically active theatre company, Odets demonstrates his style of poetic realism, but he also makes use of short scenes, characters speaking in everyday language, and showing the mechanics of transitions. All of these are common to the Epic Theatre form. One moment acknowledges the importance of the Alienation Effect when the cast steps downstage fists in the air and in unison shout "Strike!" This breaking of the fourth wall in a call to action is right out of the Epic Theater playbook.

Many of these ideas were developed as a response to the naturalism and psychological realism present in the work of Stanislavski. The irony here is in the similarities of approach. Stanislavski's focus on the actor's inner life had the pursuit of an objective at its core. Though there was much emotion generated by the actor's authentic portrayals of real people, the emotion was never a part of the actor's concern. Focus on the verb or action and the receiver of the action were at the center of the approach. The desired response, often an emotional one, was again beyond the control of the actor.

What was true of both Stanislavski's method and Brecht's Alienation Effect was that they both lead eventually to an emotional response. Brecht's focus on action and thought often made for some very emotional theatre. Self-sacrifice and courage are made more moving when the actor is not trying to play these things. The person walking down the street and seeing another walking out into traffic acts to save that person by pulling them back to the curb, not because they are trying to create an emotional moment, but because they are instinctively trying to save a life. In that instant the emotional content of the event is not the focus. The action of saving someone from harm overrides any other thoughts in the moment.

How then do we understand the implications of these ideas when considering the question of distance? In terms of the model, Brecht's work can be seen as an example of over distance. The focus is on thought rather than emotion. For actors this is known as being in your head. An actor works constantly to get out of their head to be present in the moment. Too much thinking can lead to some very wooden acting. The idea that you can't think and act at the same time is important to the serious student of acting. It is a useful concept for the communicator as well.

THE DISTANCE MODEL AT WORK IN STAR TREK

For those of you in the old school perhaps an analogy can be made using the model of the original Star Trek TV series. If we look at the main characters as a paradigm for the distance model it may be said that Spock is a classic example of over-distancing. He spends a good deal of time in his head. His response to life's challenges is to use logic and to think his way through any situation no matter how harrowing it may be and if you've ever seen an episode of that series, you know it could get pretty harrowing at times. Spock was half human, half Vulcan and though he kept his cool most of the time he was constantly engaged in an inner struggle. His default was usually to the Vulcan side though and he often used his head to ensure his survival and the survival of others.

This was of course in direct contrast to the more emotional and iconoclastic James T Kirk, the ship's captain, and under-distanced leader of the Enterprise. Kirk's responses were almost always born out of his intuition. How he felt about a situation led him to act and often act before thinking. Spock's reliance on his intellect and the use of logic over emotion was a chaser to Kirk's impulsive actions driven by emotion. It is important to take notice of how this unlikely pairing of two very different personalities not only made for a very engaging and enduring television show, but also gave us a great example of how distance works in the roles we all play.

Having these two working together at opposite poles of the distance model provides us with not only a three-dimensional look at the embodiment of the two extremes, but also a working view of how these two influenced each other to achieve, at times, the aesthetic distance that was necessary for their survival. Together these two men often found balance, one encouraging the other to stretch beyond their individual nature to fulfill the requirements of a role outside of their personal selves.

Where Spock could be rigid in his logical understanding of the world, Kirk prompted him to consider the feelings of others before making a judgment about them. Of course, Kirk's feelings were often over the top and tended to cloud his own judgments. This is where Spock's intellectual approach served to temper the reactionary Kirk and quite often literally save him from himself. Together they made each other whole. The logical thinker and the intuitive feeler made for a very effective and successful team.

It should be said that each of these men paid a price for their nature. Kirk, because of his capacity for empathy towards others, was able to find some intimacy even if it was a little forced by the conventions of television mores of the late 1960s. But he also paid an emotional price for his involvement and had to suffer through the pain of heartbreak or loss. Spock on the other hand lived a mostly solitary life. His work put him in touch with others, but he was unable to tolerate any real level of intimacy. That human part of him was buried but was in some very rare moments brought to the surface by his capacity to care for others by intellectually understanding their plight as humans. Kirk was flawed by his over emotional self, but that same example prompted humanity in Spock to be revealed at times if only for an instant.

Okay. So now my nerd is showing. So be it. The point is made. Both characters were men of science, albeit fictional. But what they tell us about the balance between the head and the heart is important. Spock brought the intellect and Kirk brought the passion. By relying on each other's strengths and compensating for their weaknesses, together they formed a fully functioning team, each gaining insights from each other and applying those to the world around them. Together they found aesthetic distance and by understanding this concept and using the techniques of working in role, you can too.

CHAPTER 9

The Importance of Rehearsal

Abstract As a final entry for the section on practice, the rehearsal process is described and considered for its importance in preparation. The phases of rehearsal are outlined and discussed in relation to the process for public speaking. Focus is also given to how preparation through rehearsal can address feelings of anxiety in the speaker.

Keywords Rehearsal • Practical games • Exercises

PROCESS ORIENTATION AND PREPAREDNESS

In the previous chapter I used Kurt Vonnegut's story, *Who Am I This Time?*, as one example to illustrate the idea of working in-role. Perhaps I thought of him because of a lecture I attended many years ago by the famous author himself. As a young man I had read all his books and was very excited to see him in person. Aside from his work as a prolific writer, it was soon apparent in this lecture that he was a very accomplished speaker as well. Understated, but intense, he spoke clearly and specifically about many things concerning his work, art, and politics.

The overall feeling I got from listening to him was that he was in conversation with the audience. Yes, he did most of the talking, but he was also in constant contact with us. He really saw us, really connected to that room full of people in a personal way. He also seemed to be impeccably prepared to share his thoughts with us. He had obviously done this many times before, but although the content was well tread, he seemed to be speaking to us and only to us for the first time.

© The Author(s), under exclusive license to Springer Nature
Switzerland AG 2024
D. Dannenfelser, *The Art of Effective Science Communication*,
Palgrave Practical Guides in Communication,
https://doi.org/10.1007/978-3-031-57030-8_9

How was it that Vonnegut could achieve such a relationship with his audience? The key I found was in something he said in response to a question from a member of the group. This person asked, "how do you become a successful writer?" Vonnegut's answer has stuck with me for all this time. He said, "the world is full of very talented writers, but there are only a very few who are willing to sit alone in a room for many hours each day and actually do it. That's what makes me a successful writer. I'm willing to sit in that room for all that time and do it."

Commitment to the process. That was the simple and valuable lesson I learned from his response. You must put in the time and do the work. Talent will only get you so far. Vonnegut was talking about his writing process at that moment, but he was also referring to his work as a speaker on behalf of his process. His talk with us was well rehearsed and he had made clear and specific choices about what it was he wanted to say to us. He knew his speech so well he could forget it and just be present as he shared his thoughts with us.

Vonnegut had spent a lifetime developing his style of writing and speaking, but a significant part of that development he owed to his commitment to a work ethic, an honest day's work. I have thought about that night many times over the years as I pursued a career of my own as a playwright, director, and teacher. In all these roles I have sought to be as present and connected as Vonnegut was to do the work that was necessary to reach my goals.

Preparation Is the Key

Preparation is the key when you set out to share your work with an audience and process is connected to preparedness. The great acting teacher and director Constantine Stanislavski wrote a book entitled *An Actor Prepares*. In this essential text he covers many aspects of his "system" for actor training. Though he speaks of the work of an actor, all his teachings are easily undone if the actor does not pay attention to the glue that holds all his concepts together, the actor's willingness to prepare.

As we have said earlier, acting is doing and doing truthfully under imaginary circumstances. When applying the work of an actor to the task of sharing a speech in the field of science there are several things to consider in your preparation. You first must be sure to have all your work on your speech in place. You start by identifying a central idea. Once you know what that is you carefully break down that idea and analyze it. You make choices about objectives and use storytelling and metaphor to help to illustrate that central idea. All this hard work can be undone by a lack of preparation when it comes time to share your speech.

Think about the work of a conductor for a symphony. When you attend a concert, you are a witness to the final product, a group of highly trained musicians working in collaboration to create something of beauty and grace. Not a few of us have probably watched the conductor standing before that gathering of artists waving their hands in the air, pointing to a particular section of the orchestra at times or whipping a baton around at others. "I could do that," you may have thought to yourself. "What kind of job is that?"

What you may not realize is that every one of those signals the conductor sends to the members of the orchestra represents weeks of rehearsal process in a room in the basement of the concert hall. The shorthand of communication we see between the conductor and the orchestra during a performance is the icing on the cake. Endless discussions about how loud or soft a measure of music should be played, or how fast or slowly, or of the emotional essence of one passage and the opposite requirements of another are all a matter of process lived through in the rehearsal hall. Spontaneity isn't an accident, or some mysterious phenomenon reserved for those of a special talent. If you remember the joke shared by my improve company to prepare something spontaneous, you may now see that there is some wisdom there. Preparation frees you to be open to whatever happens in the moment. Spontaneity is the result of a lot of hard work. The performance, what we see, is gravy.

REHEARSE, REHEARSE, REHEARSE

What is the benefit of time when undertaking a new task? There are few activities that require a skill of some kind that can be accomplished with little or no practice. It is necessary to devote time to the development of those skills. In the theatre we call this rehearsal. Just as an athlete must attend practice to work on the skills necessary to play their respective sport, the actor must attend rehearsal to develop the skills necessary to play a role. Introduction of new ideas, actions, or processes of art requires repetition. We engage in the practice of rehearsal to challenge ourselves to gain the knowledge and skill necessary to authentically portray a character on stage.

But a curious thing happens when people face the challenge of public speaking. For some reason many believe that they should innately understand how to do this, and they must be good at it immediately. Public Speaking requires the development of certain skills and to do that you have to first study the form and then rehearse for the performance. Because it *is* a performance although you may not think of it that way and every great performance requires hours and hours of rehearsal. It takes a lot of time and effort to make something look effortless.

PHASES OF REHEARSAL

In his book *A Sense of Direction*, Bill Ball outlines the phases of rehearsal (Ball 1984). I have taken his work on this and adapted it to the speech preparation process. What is significant about each of these phases is the progression of preparedness that becomes apparent as you go through each of them. Each one has its purpose and should be given the proper amount of attention. The phases are also sequential and the movement on to one phase is dependent upon completion of the previous one. Here is my adapted sequence with some commentary:

The Rehearsal Process

FOR SPEECHES

- **THE SIT-DOWN READING & DISCUSSION (TABLE WORK)**
  The focus here is on the story and making sense of it all. Do you have a clear and specific Central Idea? After having discovered the Beats and Objectives in the writing process you do some further process during Table Work to refine these choices. You will also develop an outline for further process on your speech here. Specific attention in discussions is given to the audience. Who are you speaking to and what is the central idea you want to express to them in particular?

- **BLOCKING REHEARSALS**
  The initial movement you will use as you share your speech. This is tied to what you want to communicate in each beat (objectives). Get on your feet and adjust your movements for optimum impact and efficiency on stage.

- **OFF-BOOK REHEARSALS**
  The painfully essential time when you struggle to free yourself from the text of your speech and claim ownership of the material by memorizing it completely. You must be Off-Book before you can move on to the next step.

- **WORKING REHEARSALS**
  The most creative time for you. Now you work to sharpen intentions by testing objectives and deepening understanding. New ideas come out of the unexpected and spontaneous moments of improvisation where you are free to discover through "play."

- **RUN-THROUGHS**
  Bringing together all the pieces, beat by beat. A partner can give you notes towards adjustment and clarification of all the points related to your central idea and your choice of objectives. The goal is to keep going to give you a sense of continuity and seamlessness.

- **SPECIAL SCENES**
  Time is given to those parts of your speech that require more attention. They become more apparent after having run-through the speech a few times. Make sure to allow time in your schedule for this specific polishing. Details, details, details.

- **DRESS REHEARSAL**
  Another complete and uninterrupted run-through of the speech. Again, the goal is to keep going to give you a sense of continuity and seamlessness.

- **OPENING NIGHT**
  Break a leg!

You may have surmised from the phases described above that the time required to pass through each of these is significant. The actor typically commits to weeks and sometimes months of rehearsal to prepare for a performance. Mastering each phase is an organic process. You can't skip around in these building blocks to performance.

There is another element of time that needs to be taken seriously in this process, the time that is required between each phase of rehearsal. The creative process is akin to the growth cycles of all living things. We all remember those awkward years of self-discovery in our adolescence. Though painful at times, each phase of our development as fully human people took time. Some of us may have felt stuck in a particular phase before we were able to move on to the next. That feeling of being stuck is also a part of the growth process. The urge we felt to move on could only have come from living through the experience at each phase.

Growth is difficult, challenging, and painful at times, but necessary for the flower of creation to bloom. In life we go through these stages to become ourselves. In the imaginary life we go through these stages to discover our creative selves. The time spent germinating between each phase is significant and necessary. You can't rush the process so plan and give yourself the time to live through each one of the phases.

A good example of this concept can be found in the third phase of rehearsal. The off-book is a difficult and necessary phase. The purpose of the off-book rehearsal is simple, to get off-book. It is a messy and challenging process. When it comes time to share your speech, all text should be memorized. If you are still thinking about the text of your speech, you are not prepared. By that time all of the text should have been turned into action through objective.

When Polonius asks Hamlet, "What do you read, my lord?" Hamlet's answer is, "Words, words, words." This passage is descriptive of this process. The struggle here is to go beyond the words and turn them into actions. To simply memorize a speech is not enough. The actor has spent much of the time at the table with the script, not unlike the musician with the score, to discover the meaning of the words in relation to a specific and doable objective motivated by the relationship with their partner. The proof is in the doing.

Anxiety, Preparation, and the Role of Homework

To be unprepared is to be anxious. Much of what we experience as nervousness about a speech comes from that feeling of fraudulence that accompanies a lack of preparation. The response to this feeling is simple, prepare. And yes, I know, simple ain't easy, but it is simple. You just have to do it. Work hard in rehearsal and then leave all that work at home and be present in the moment of performance.

As I have mentioned above, both Vonnegut and Stanislavski speak to us about the value of preparation. The rehearsal process gives us a structure for this, and we allow ourselves to go through each phase to move on to the next with a sense of completion. The periods in-between rehearsals are a time for homework. You commit the time to the discoveries you make and then to the actions they prompt in you as a response. If you do this, you will be ready for performance when you must forget about your homework and be present in the moment.

PRACTICAL GAMES AND EXERCISES

Predator/Prey (Conflict)

Setup and Instructions
- Two players are blindfolded.
- Both enter a circle created by the other players.
- A simple weapon (feather or rolled up paper) is placed somewhere inside the circle.
- Blindfolded players begin to search for the weapon. First to find it is the predator and the other the prey. Leader calls "weapon up" when the weapon is found.
- Predator hunts the prey and must touch him or her with the weapon.
- Players around the circle protect the inside players from harm and keep them inside the circle by gently guiding them with touch. This is a non-verbal interaction.

Extensions (More Non-verbal Interactions)
- Players around the circle help the predator.
- Players around the circle help the prey.
- Players around the circle help the player of their choosing.

Essence of Discussion
This exercise creates a lot of tension and excitement in the members of the circle. All become invested in the hunt on one side or the other. Careful commitment to the non-verbal rule heightens the tension in the circle. The concept of stakes are introduced by the extensions which prompt the players in the circle to make a choice. This becomes an important point when related to the building of a story for your speech. What is at stake? Why should anyone but me care about this research?

Story in a Circle

Setup and Instructions
- While seated in a circle, the leader starts a story, i.e.: "Once Upon A Time."

- Each player adds to the story with a statement that begins with: "Yes and…"
- The story ends when the leader says "The End."

Essence of Discussion
It takes everyone involved to be successful. Everyone has a role to play, and the group must depend upon each other. A doctor can't practice without the patient. They are the reason you are there in the first place. Take your cues from them.

I'm Thinking of a Story

Setup and Instructions
- Instructor asks players to try to guess what the story is.
- Instructor answers only with yes, no or maybe.
- Yes to any question ending in a vowel.
- No to any question ending in a consonant.
- Maybe to any question ending in a Y.
- Adjust if too many nos.

Essence of Discussion
Your focus is on discovering the story without fear of approval or disapproval. By taking any prior knowledge of the story out of the mix, the players are forced to focus on doing as they create the story they don't know. This misdirection frees the players from judgement about the "right" story. The forming of the story is the point, not the quality of it, good or bad as it may be judged to be.

Proverbs

Setup and Instructions
- Group stands in a circle.
- One player says a word. (First impulse)
- Another player adds another word. (First impulse)
- When the group feels the proverb is finished, all tap fingers together and say, "yes, yes, yes."

Essence of Discussion
This fun and simple game sparks the imagination and prompts the use of imagery that is evocative and economical. Again, the unknown is the activating element. There is no expectation as to what the proverb should be. It is what we make it in the moment. When applied to a speech, proverbs can get to the essence of an idea in less prosaic and more engaging way.

Exposure (Spolin 1963)

Setup and Instructions
- Half the group stands in a row facing the rest of the group who are sitting.
- As they stand the sitting group observes for about two minutes.
- The two groups switch places.
- The second group just stands as the first group had for a minute or so.
- The second group then is instructed to count objects in the room, chairs, tables, etc.
- The first group observes and notes any changes.
- The first group then stands again. At first, they only stand. After a moment they too begin counting.

Essence for Discussion
What is the meaning of objective? How do you use the reality of doing? This exercise focuses on the actual pursuit of an objective through doing. The group first experiences the absence of a clear and specific objective and then is guided towards a simple and doable one. The difference in how the players feel when standing without a goal and when they are involved in counting the chairs is made clear. Having no clear objective is uncomfortable. The reality of doing creates ease and a lack of self-consciousness.

Gibberish Demonstration (Spolin 1963)

Setup and Instructions
- One player stands at the front of the room with a chair.
- That player calls upon another to join them.
- The first player then demonstrates some activity to the second player regarding the chair using only nonsense syllables or blah, blah as language.
- The second player completes the activity.
- The second player now chooses another player, and the process is repeated.

Essence for Discussion
The POC here is to communicate without the benefit of language. Demonstration or doing is used to replace the more intellectual process of speaking. When sharing a speech, language is only one means towards communication and can often complicate the subject matter unnecessarily. Showing rather than telling can be a useful point of view for the public speaker.

Three Views of You

Setup and Instructions
- All sit in a circle.
- First round: Go around the circle and say one thing you observe about somebody you see.
- Second round: Go around the circle again and say what you assume about the same person based on what you observed.
- Third round: Go around the circle again and say what you imagine about the same person based on what you assumed.

Essence of Discussion
The use of observation, assumption, and then imagination tells stories about ourselves and the people in the world around us and the roles we play. These can be adjusted through discussion to both accept and reject certain ideas and judgements we make about each other. Adjustment is the key here. Some of what is assumed and imagined may be true. Adjustment incorporates those truths while also clarifying them to create a more accurate picture of a person. The possible roles that are explored can then be considered as possible realities for the person as they consider how they are viewed by others.

Changing Places

Setup and Instructions
- Each player gets a number.
- All travel space until Instructor calls freeze.
- # 1 initiates movement from this point on.
- #2 follows #1, #3 follows #2, and so on. Each fills the other's space exactly.
- Each player cannot move until the previous player has moved.

Essence of Discussion
Simply taking on the physical way another inhabits a space can lead to insights into that person's view of themselves and the world in which they live. When you try to move as if you were another person, you get an idea of how it is to be in another's shoes. This exercise opens the possibility to see things from another's perspective using the physical. As the body goes so too can the mind.

Quote Sculptures

Setup and Instructions
- Two teams
- One team calls out a quote that has personal meaning to them.

- The other team creates a sculpture using their bodies that is a visual representation of the idea of the quote.
- Switch and repeat.
- Move into a circle and have whole group create one sculpture together.

Extensions
The Real Image

- One steps out and adjusts the sculpture according to opinion on accuracy.
- After several adjustments the sculpture is agreed upon (consensus) as being the real representation of the idea of the quote.

The Ideal Image

- Group steps back and re-creates the sculpture based on an ideal of the idea of the quote (a perfect world view).

The Image of the possible transition

- Return to original Real Image
- Move from the Real to the Ideal in slow motion together.

Essence of Discussion
Once again, adjustment is used to consider ideas from a different point of view. The connection of the body and mind through physical representation is powerful. Trying out different views that include an ideal image of an idea creates flexibility when that ideal is then adjusted to incorporate what is wished for into what can be. This is a reality check that uses extremes to arrive at a balanced view of yourself and the world in which you live.

PART III

Performance

CHAPTER 10

Relationship to the Audience

Abstract Knowing your audience is the first step leading to performance for the public speaker. In this section, performance is discussed as the natural culmination for all the theory and practice explored thus far in the book. Again, preparation is the key and myths about judgment and failure are considered as they are addressed by the speaker's attention to being prepared. A distinction is also made between sharing a speech rather than presenting one. The concept of sharing is another step towards a confident and authentic performance.

Keyword Audience relationship

KNOWING YOUR AUDIENCE

Is it possible to really know an audience? Probably not, but you can determine certain things about them that will help you as you share your speech with them. You start by asking yourself simply, who are they? What is the context for this event that presumably has you at its center? It is in your interest to learn as much as you can about the audience before you share your speech with them. How are they like you? How are they different from you?

My friend and colleague Dr. Nick Ponzio liked to use the comparison of apples to oranges. The image of a speaker talking about apples to an audience whose preoccupations were with oranges was a simple but effective analogy. If you know that oranges are the central issue for an audience, you would do well to couch your remarks with some regard for one fruit rather than another.

A little bit of research is in order when you have the benefit of knowing who you will be speaking to in advance. Websites have a good deal of information that can help in this effort but making the most of the initial contact with a person representing the group who will be attending your speech is essential. In a sense you are engaged in an interview process when you first contact a group or their representatives. Make the most of that moment and recognize that you are not only the interviewee, but also the interviewer. Ask a lot of questions. Try to determine what the goals are for this gathering.

When I am asked to run a workshop or speak to a group of some kind, I first try to determine what it is they want. Sometimes they don't really know what they want. In that case my questions help us both to clarify the reason for the event. Simple questions about the history of the organization and their past efforts and practices can yield a more specific determination as to what the content or the style of the event should be.

Just as you prepare a speech with a central idea in mind and make choices about objectives based on that spine, the audience for the speech plays a role in the determination of how that speech will go. Often there is a challenge or problem of some kind that has caused the group to reach out for help or guidance. This is a great place to start. Like the dramatic structure of your speech itself, the performance of your speech addresses some desire or want in the group. You do well to ascertain what that is before you show up to speak.

Preparation and the Teachable Moment

The common myth about public speaking and the source of the dread associated with it is this idea that the audience is out to get you. While it is true that an audience of fellow scientist can be demanding, it is important to understand from where those demands come. As mentioned before, the scientific method sets a standard. Your speech about your research must be evidence-based, your conclusions clear and supportable. Many of your fears about a challenging audience can be soothed if you are sure to follow a professional process for the formation of your ideas.

Preparation again is the key here. If you have in fact done your work in an authentic process, then you are already on your way to a solid performance of your speech. Preparation allows you to speak with authority and without apology. Does that mean you may not have missed something? No. Your honest attempt to follow a professional standard does not guarantee that your findings will not be challenged. But if you have done your very best to prepare a supportable argument for your research, then any challenges to that should be met with an open mind. There may be a subtle aspect that you missed or a new finding that adds a different color to your work not discovered by you initially. These kinds of challenges should be met with openness and taken seriously. You may learn something on this occasion that you will incorporate into future revisions of your speech. That becomes a teachable moment for all concerned and not a reason to panic for you.

The Mythology of Judgment

But what about the idea that persists about an audience's general hostility towards a speaker? Can it be said that an audience is out for blood whenever a speaker attempts to share a speech with them? I don't believe so. If you can view your audience as just as serious about the work as you are then you can begin to put this relationship into perspective.

For example, the relationship between the public speaker and an audience can be compared to the relationship that exists between an actor and director in an audition process. The common misunderstanding of the audition process has been mythologized in the theater world for a long time. In this view, the director is a tyrant who wants nothing more than to victimize the actor and make them out to be a fool when they step into that audition space. But don't believe it.

As a director of theater, myself, I know that nothing could be further from the truth. Why would I go to all the time and trouble to hold auditions to fill a role I need for a production to simply make this an opportunity to belittle every actor who bothers to show up? What kind of fool do *I* have to be to shoot myself in the foot that way? When I hold auditions, I'm hoping to see actors at their best. If they come in and give me what I'm looking for then my job is done. I want to celebrate actors when they come out for an audition. Putting them down would only reveal how little I think of myself wishing to berate someone to make myself feel better about my own feelings of insecurity.

Does this happen? Yes, it does, but it is the exception and not the rule. Most directors are smart and secure. They know that a comfortable actor is a relaxed actor, and a relaxed actor is going to give you their best work which helps you to do your own. I do my best to put actors at ease when they audition. I acknowledge any nerves they are experiencing by talking about that and letting them know they are among friends. I tell them we are in this together and I am rooting for them. I am also clear that this is not personal. I'm interested in your work, your skills as an actor. I'm not interested in judging you as a person. Show me what you can do and if you can have some fun doing it.

For the most part, the same can be said about your relationship to an audience. Together you enter into an agreement. They have come here today to hear what you have to say and to learn something they maybe didn't know before you shared your speech with them. In short, the audience wants you to succeed because if you do, they get what they came here for, an illumination of a certain subject by a qualified speaker. They are on your side. They are your partner and not your judge.

Sharing Your Speech vs. Presenting Your Speech

For the student of Science Communication, there is much to unpack here. First off, what is the purpose of the task to share your speech? Is it to explain what you do to others? How will you proceed and what is your attitude about sharing your work with others?

I have used the word sharing very intentionally throughout this book. When we think of speaking publicly, we often think of this experience as a presentation. We are expected to stand up in front of an audience of our peers or strangers and present our research to them. The "us" versus "them" relationship that is baked into this point of view of public speaking can lead to not a little amount of anxiety for the speaker. The pressure is on you to deliver a speech and to gain the interest of the audience.

On the other hand, to share your speech with that same audience is a very different mind-set. This idea is connected to my comments on GOS or generosity of spirit in an earlier chapter. The idea that we are in this together creates a different dynamic in the room. I have done the work to prepare this speech for you and with you in mind. I am now going to share this speech in a way that keeps you in mind and to some extent even guides the way I share it. In this approach the audience is an active participant in the sharing of the speech. The audience becomes your partner rather than your judge and if you remember our study of the actor's process, your partner is the most important person on stage. Think of sharing your speech with a partner rather than presenting it to an audience.

CHAPTER 11

Sharing Your Speech

Abstract The idea of sharing is considered in more detail as well as a look at the person in the role of speaker. This chapter is also a review of how the creative process leads to a positive and productive sharing of a speech. Concepts such as keeping your homework at home and having a clear and specific objective related to a partner are also reviewed here.

Keywords Speech sharing • Performance • Practical games • Exercises

WHO ARE YOU?

Who then are you when you set out to share a speech with an audience? The answer is you are many people. You have written the speech like a playwright, you have made choices about your objectives like an actor, you have imagined the style of the speech like a designer, and you have analyzed, broken down, and made choices about the approach to the overall experience for the audience like a director. All this work has been guided by a point of view that was determined by your commitment to a clear and specific central idea. The roles you have played up to this point have been in preparation for the final role in the process, that of speaker. Like the actor who plays the role of a character in a performance of a play, you now play the role of speaker in the performance of your speech.

SHARING IS A PERFORMANCE

When you share your speech, you are creating a performance. The give and take that happens between you and the audience is a performance that you form together. That being true it is necessary to be sure that each of your objectives have a test in the other, your partner, the audience. You should be able to see and feel a response on the part of the audience as you move through your speech beat by beat. The *how* of your performance as speaker is directly connected to work you did in preparation for your speech.

After all the analysis and identifying of transitions leading to selections of beats and objectives you must now decide upon tactics to see those objectives come to life during your sharing of the speech, the performance. You have broken down the speech. Now go back and look at the manner in which you are expressing your ideas related to each objective. You are trying to develop an approach to your speech that gives your audience access to what you have to say.

Ask yourself, how will I *awaken* (verb) my audience (receiver) to their sense of outrage or concern about my central idea (desired response)? Go back and consider the tactics you are using to express your central idea. What are your metaphors? What is the emotional essence? What is the premise of your story related to your central idea? An objective is theoretical until you put it to work. How will you go about connecting to your audience? The objective will be your guide.

Remember just as an actor cannot play an emotion on stage, you cannot play the emotional content of your speech. You must pursue the action, the verb in relation to your audience. Really taking them in and responding to their responses to you in the moment. This is about being present and there is no better way to be present than to get busy pursuing the actions built into your objectives.

The *how* looks different for every actor on stage and for every speaker sharing a speech. But each of these is guided by a strong, clear objective and a plan of action containing tactics related to reaching that desired response. The way you speak, your inflection, the way you move or gesture, the intensity of your performance, all are results of your commitment to clear and specific objectives guided by what you want. When you know exactly what those objectives are you will begin to form a plan for how to get what you want.

HOMEWORK STAYS AT HOME

I have mentioned this idea before as a part of your preparation in rehearsal. Now let's look at this concept in relation to the job of the actor and ultimately to your role as a speaker.

Why does an actor learn their lines? The answer is to forget them. An actor doesn't bring their homework onto stage. All the work an actor does to rehearse brings them to the moment when they must be present on stage to plant their feet, look their fellow actor in the eye, and tell the truth. An actor can't be present if they are still thinking about their lines or blocking or even their objectives during a performance.

By the time an actor is on stage all the table work they did, all the blocking they discovered, and all the objective choices they made should have been transferred from their head into their body, from theory into practice through action.

What then is the job of the non-actor sharing a speech? The same. Plant your feet, look your audience in the eye, and tell the truth. If you follow the rehearsal process and really commit to it, you will be ready and prepared. You won't have time to be anxious. You will be too busy doing what you came there to do.

Going beyond memorizing text and having discovered what your objectives are, when it comes time to share your speech, if you are still referring to those objectives for each beat as ideas instead of playing them as actions, you are not prepared. By the time you are sharing/performing your speech you should be *doing* what you decided you *wanted* to do and not thinking about what that is. All your objectives should have been transferred into action. Don't bring your homework onto stage.

The Test Is in the Other

Once you have done all your preparation, your focus during the performance of the speech is on sharing your work with the other, your audience (Partner) in a give and take relationship.

The audience will give you responses that tell you when they are with you or when you have lost them. You need to be aware of this to adjust in the moment. But to be clear, this is in relation to your speech and the central idea that is served by your objectives. In other words, it's about what you have to say not about who you are. If some people are going to judge you, you will have no control over that.

Your focus is on your work and on effectively communicating the central idea of that work. Your concern with the audience is related to whether they are getting what you have to say and nothing else, period. Being "liked" is not an objective. Hopefully, if you do your preparation, plant your feet, and tell the truth the audience will like you, but it will be because of what you have shared that enlightened them and not because of who you are. In this way, the focus is on the other and not on yourself, but ultimately the real focus is on the work you have come to share.

Practical Games and Exercises

How's it Like Me? (Austin, Boyd 1973)

Set Up and Instructions
- One player volunteers to leave the room.
- The remaining players choose an object in the room that somehow reminds them of the missing player.
- The missing player returns and asks three other players one at a time to describe a quality of the object that reminds them of themselves without naming the object or giving it away.
- The missing player tries to guess what the object is based on this information.

Essence of Discussion
This form of shedding light connects a person to an object by focusing on specific qualities. What makes someone who they are? Going beyond the surface to look at how a person behaves and to make certain determinations about what they value or care about lead to insight. Knowing your audience is important. Seeing them as people like you is essential.

Dumbo Crambo (Two Groups) (Austin, Boyd 1973)

Set Up and Instructions
- First Group exits room.
- Second Group chooses a verb to act out.
- First Group returns and are given a nonsense word that rhymes with verb chosen by Second Group.
- First Group acts out the verb they think is correct.
- Second Group guesses what verb is being acted out.
- If First Group is correct the Second Group says YES it is VERB and applaud.
- If First Group is incorrect the Second Group says NO it is not VERB and shake heads.

Essence of Discussion
The *how* is important when pursuing an objective. To form the objective is one thing, but how will you actually pursue what you say you want? Tactics must be considered as a means towards that end. Public speaking like acting is about doing. Choose an approach to every objective you form and really do it.

PART IV

The Way Forward

CHAPTER 12

Perceptions, Misconceptions, and Final Thoughts

Abstract The final section of the book looks to the future for the effective communicator. Here, traditional ideas about public speaking and the connection between the actor's process and that of many others are revisited. Common apprehensions and misunderstandings about the art and science of public speaking are explored and hopefully dispelled in this chapter. A more specific look at how recent movements in STEM (Science, Technology, Engineering, and Math) education have been adjusted to include the Arts and change the acronym to STEAM are also found in this chapter.

Lastly, this chapter summarizes the book and encourages the reader to make their own choices about the kind of communicator they wish to be. In a return to the introduction, the concept of embracing the introvert is repeated here as support for the effective communicator in the challenges ahead.

Keywords Perceptions • Misconceptions

FORSAKING YOUR SCIENCE

I've touched on this idea earlier, but there is more to say on this misconception of what our work as scientists and effective communicators is all about. There is a very real fear that exists in the groups of scientists I meet who have an interest in communication. The fear is that by making their work accessible they will somehow have to dumb it all down for a general audience. For some this is tantamount to not speaking honestly about their research at all. That is not necessarily true, though it is understandable.

When you are faced with the task of explaining what you do to others who have no experience in your field you must try to think about your work in terms that may be understood by them. As a theater professional I have been challenged in the past with the task of explaining what I do to others. It isn't easy to sum up my work in a short conversation at a party, especially because I have developed a few hybrid incarnations of it.

I am an educator/playwright/director/communications specialist. Any one of these would prompt a lengthy discussion of what that means and because I am an introvert it was always my hope to keep these kinds of discussions to a minimum. I would often defer to the work of my partner in the conversation preferring to hear from them. I sometimes minimized it all for strangers by just saying, I'm a teacher. But even that would prompt the question, "what do you teach?" And so it goes.

The point is, my career has been complicated and all roads did not always lead to Rome, to use a proverb, where my work was concerned. But if I wanted to connect to others, and I did, then I had to find a way to speak about my work succinctly and accurately. Yes, I could have just lied and said I was a plumber, but why should I have to miss out on true connection to others just because my work was out of the ordinary?

You can see where I'm going with this. Scientists are often faced with this dilemma. How to make what I do understood by non-scientist but also remain true to who I am. How to keep my passion for my research and share that passion with another who has no idea what it means. Do I have to just be a plumber and leave it at that? By the way I have nothing but respect for plumbers. I'm sure they too have avoided talking about the intricacies of the trade to keep the conversation going.

How then do any of us share our work with the uninitiated, but also remain true to that work? Teachers, plumbers, and scientists all spend a lot of time perfecting their crafts. Do we have to just forget about all that effort when we try to communicate what we do to others? Not really.

The scientist must be accurate and evidenced based when speaking about their research, but they also need to be understood at times by non-scientists. People who have influence and power over the funding necessary to continue that research aren't necessarily versed in it. Politicians, business professionals, and members of the media who can champion your work or put an end to it must have *access* to a working definition of what you do. That word, access is important. Making your work accessible is necessary for its survival. The public may not understand what you do, but they often have control over whether you do it or not. It is in your interest to make your science understood by anyone who is willing to listen to what you have to say about it.

THE CHALLENGES TO EFFECTIVE COMMUNICATION

As I have said throughout this book, the task of public speaking can be daunting. There are many reasons for this, but a fear of rejection is at the center of most of them. The ridicule a scientist may face publicly when their research is challenged is something everyone would like to avoid, but if you follow a legitimate process in preparing your speech the same as you would for conducting an experiment, then you will do much to allay your fears before you find yourself speaking to an audience.

Taking a positive point of view of your work and of the public speaking event is a good start. Remember like the actor you can't pursue a negative on stage. Saying yes and justifying that choice is full of action and will keep you moving forward no matter what happens. But sometimes this is not so easy.

The feeling of being an imposter when speaking about your work is common. Who am I to be telling others about this very important aspect of science? Why should anyone listen to me about it? The answer to this may be, why not you? Those feelings of fraudulence come from an incomplete preparation process. To endow your sharing of your work with too much power is a response often to being unprepared. So, what do you do? You do the work. You prepare your speech carefully and systematically as we have explored and then you rehearse the performance of your speech in earnest. You put in the time and the effort to be ready. Just like an actor can't play a role they don't understand, you must be thoroughly versed in your understanding of the work you do. You first study the science inside and out and then study the process by which you share that science with the same kind of intensity and commitment. You are not a fake, an imposter when you can say, I earned this moment and I have the sweat to prove it.

I have had students in my classes question the validity of this process referring to the same issues we are talking about here. Is this really my research if I talk about it in a way that can be understood by non-scientists? Am I just dumbing it down for the sake of being understood, but losing the thread of what my work is about? It is not surprising to discover that these students are often the same ones who are unprepared to share their speeches in class. They put a lot of energy into complaining about the efficacy of the process and not enough time working through it. They are avoiding the work to confirm their own apprehensions. If you want to feel like you have a right to talk about your work, put in the work it takes to earn that right.

Soft Skills vs Hard Skills and the Myth of Talent

Art is a soft skill. Science requires the development of hard skills. In other words, the common understanding of the work of an artist is that you don't have to work that hard to achieve a certain level of success. This goes hand in hand with the myth of talent. People who do the creative work of art are just magically prepared for that work because of a God-given talent. They just know and understand how to be a great musician, painter, actor, etc. I can't tell you how many times I've heard this kind of bosh coming from people after they have just had a significant experience of art. "That person is so talented. I could never do something like that." They mean it as a compliment, but really it is a furthering of the myth that some people can just be creative artists without trying very hard at it.

Likewise, the scientist is seen as a rarefied being who has the power to understand difficult concepts and technical information that cannot be understood by commoners. "I will never be able to understand how that person does what they do." What happens because of these long-held beliefs and attitudes is that both the artist and the scientist are kept at arm's length. They occupy certain places in society that cannot be shared by or with the public. Better we just worship or disdain them for it than try to understand the roles they play and how we are also a part of that experience as we all share a fragile and beautiful existence as human beings.

STEM AND STEAM

The bridge to understanding is communication of course but you can't just expect that will happen. We must work hard together to bring our audiences into our worlds. The artist is at their best when the work they do reflects the experience of their fellow human beings or as Hamlet says in his advice to the players, "to hold as 'twere the mirror up to nature...." The same can be said of the role of the scientist in society. The STEM (Science, Technology, Engineering, and Mathematics) programming that has gotten much attention in the past few years is incomplete if it doesn't include a serious consideration for the role of the Arts. We should be on a quest together to change that acronym to STEAM and add the Arts into our programming. The marriage of these disciplines is stronger when consideration is given to how they are understood by the public. Those people who don't really know what it is the artist or the scientist does must be invited into our conversations and are vital to the continuation of the work of both.

THE CHOICE IS YOURS

If you are a serious student of communication then you have probably found that there are many different books like this one speaking about many different, but related approaches to this task. The flood of information in the field can be confusing and overwhelming at times, but there is a way to deal with this.

Many years ago, I worked at a Theater Institute at Oberlin College in Ohio. I was there along with several other faculty for a summer intensive in acting for prospective students. I set up my class in the usual way and entered this rigorous training program in earnest. After a few class meetings the students lobbied me for a discussion about our approach to the work. These were smart students, inquisitive and serious about their training. They wanted to know why it was that the approach to the work of an actor that I was teaching was in direct conflict with what another teacher was using just down the hall.

The students were exposed to many viewpoints that summer and were encouraged to do as much training as they could to get the most out of the experience. But this circumstance was making their experience confusing. How is it that we can attend your class in the morning and learn about one approach to acting and then attend another class in the afternoon that completely contradicts everything we learned earlier in the day? A good question and the answer I gave to them was simple and has implications for you as you go forward in your studies as a communicator.

My answer was another question, "which one works for you?" This is art. The subject matter and the approach to it is open to interpretation. When faced with a contradiction you are being given an opportunity to make a choice. In the work of an actor, when someone presents you with a limitation, you say thank you. Your limitations are your freedoms. If you ask a comedian to be funny, the proper response from a professional is, "about what? Who am I, where am I, what do I want?" Remember that? If you think of it this way then a limitation is not a limitation at all. It is an invitation to be specific about what you are doing. It also challenges you to make a choice and really look at the best choice for you.

For the student of communication, it is a good idea to explore the many different choices and then find the one that grabs you by the throat and says this is it. If you pursue this work in earnest, then you will have the experience of comparison and you will get a feeling for how one approach is affecting you as opposed to another. When that one approach hits you, you'll know it. It will just make sense to you. You'll know what to do with it, how to act in response to it. When that happens, then you will know, "this is the approach for me."

Seeking out different views on the subject of communication or any subject for that matter is a good way to keep things in perspective. One course or book on the art of communication does not a master communicator make. When the semester is over and we are wrapping up our course in Science Communication at Rutgers, I tell the students, "This is not the end, this is the beginning. Be curious. Fill yourself up with experience and then make a choice about what works best for you."

AUTHENTICITY AND EMBRACING THE INTROVERT

I began this book by declaring myself an introvert. I return to that statement now as a final thought for your work going forward as an effective communicator. As an introvert, the way forward for communication has been a challenge for me, but it is who I am. So rather than denying that reality I have learned to embrace it and remain true to my essential nature. There is a saying we have in the theater, "If you can't fix it, feature it." And so, by understanding the role I am playing as a public speaker I have been able to express myself from the aesthetic distance of that role. I have learned to feature my introverted nature and channel that point of view into a role to help to achieve the task at hand, true connection to an audience in a shared space surrounded by an assumption of good will.

There is an old joke mostly attributed to the comedian, George Burns. It goes something like this, "The key to success is sincerity. If you can fake that, you've got it made." In speaking about his own success, Burns took a more personal view. People would ask him to what he attributed his considerable acclaim and longevity as a comedian. To understand his answer, you must first know that he was married to another comedian, Gracie Allen. Burns' answer to the question was simple. He explained, and I'm paraphrasing, "Each night I'd come out and ask Gracie something like how's your brother and Gracie would talk for twenty minutes and that's how I became a star."

Focus on the other. My partner is the most important person on stage. These concepts I have been talking about have been around for a long time. I didn't invent them, but it is important to repeat them now in these final thoughts. My partner, not me, is at the center of every choice I make, every objective I pursue, everything I do as a public speaker. My partner, the audience, is the reason I am here. They keep me focused on the central idea. They let me know when I am connecting or when I've lost them. They keep me honest, and *they* hold the key to my success.

I return to Burns' joke because it brings to mind a concept I would like to mention as this book draws to an end. The concept is concerned with the idea of sincerity and for me its relationship to something else, authenticity. The joke he tells is funny because it is the opposite of the true meaning of the word, sincerity. But what is the relationship between this word and the other one, authenticity? Simply put, in my view, authenticity is about staying true to yourself, and sincerity is about sharing that true self with others. To share your work with others in a sincere way, you must first find your authentic self in it. Be you an introvert, an extrovert, or any of the myriad of nuanced personalities of people that exist on the spectrum, when you are determined to share your work with others with sincerity, try to keep it simple, plant your feet, look your partner in the eye, and tell the truth.

BIBLIOGRAPHY

Alda, Alan. 2017. *If I Understood You, Would I Have This Look on My Face?* New York: Random House. ISBN: 9780812989144.

Austin, Ruth, and Boyd, Neva L. 1973. *Handbook of Recreational Games.* New York: Dover Publications. ISBN: 0-486-23204-2.

Ball, William. 1984. *A Sense of Direction.* New York: Drama Book Publishers. ISBN: 0-89676-081-2.

Boal, Augusto, Jackson, Adrian, Trans., 1992. *Games for Actors and Non-Actors.* London: Routledge. ISBN: 0-415-06154-7

Campbell, Janice, 2023. Plot Diagram: Freytag's Pyramid - Excellence in Literature, Excellence-in-literature.com.

Carnegie, Dale. 1936. *How To Win Friends and Influence People.* New York: Simon & Schuster, Inc.. ISBN: 0-671-02703-4.

Egri, Lajos. 1960. *The Art of Dramatic Writing.* New York: Simon & Schuster.

Jung, C.G., Jaffe, Amiela, Winston, Richard, Winston, Clara, Trans., 1961. *Memories, Dreams, Reflections.* New York: Random House, Inc. ISBN: 0-679-72395-1

Landy, Robert. 1986. *Drama Therapy.* Illinois: Charles C. Thomas Pub.. ISBN: 0-398-05176-3.

Meisner, Sanford. 1987. *Longwell, Dennis, On Acting.* New York: Random House, Inc.. ISBN: 0-394-75059-4.

O'Neill, Cecily, and Lambert, Alan. 1982. *Drama Structures.* London: Hutchinson Education. ISBN: 0-09-147811-1.

Peale, Norman Vincent. 1952. *The Power of Positive Thinking.* New York: Simon & Schuster. ISBN: 978-0-7432-3480-1.

Spolin, Viola. 1963. *Improvisation for the Theater.* Illinois: Northwestern University Press. ISBN: 0-8101-1000-8.

Stanislavski, Constantine. 1936. *An Actor Prepares.* New York: Theatre Arts, Inc., Routledge. ISBN: 0-87830-983-7.

D. Dannenfelser, *The Art of Effective Science Communication*,
Palgrave Practical Guides in Communication,
https://doi.org/10.1007/978-3-031-57030-8

Vonnegut, Kurt. 1950. *Who Am I This Time from Welcome to the Monkey House*. New York: Dell Publishing Co.

Wagner, Betty Jane. 1976. *Dorothy Heathcote- Drama as a Learning Medium*. Washington, DC: National Education Assoc.. ISBN: 0-8106-1383-2.

INDEX